Science Teachers Who Draw
The *RED* Is Always There

Merrie Koester

DEEP UNIVERSITY PRESS

Deep University Online !

For updates and more resources,
Visit the Deep University website:

www.deepuniversity.net

www.deepuniversitypress.org

Copyright © 2015 by *Deep University Press*
A subsidiary of Poiesis Creations Ltd, Wisconsin, USA
Member of Independent Book Publishers Association (IBPA)

All rights reserved. Permission is granted to copy or reprint portions up to 5% of the book for noncommercial use, except they may not be posted online without written permission from the publisher.

For permissions, contact: publisher@deepuniversity.net

ISBN   978-1-939755-22-3 (Hardcover)
        978-1-939755-08-7 (Paperback)

Library of Congress Cataloging-in-Publication Data

1. Science Literacy. 2. Science and Drawing. 3. Arts-Based Educational Research. 4. Science and Art. 5. Qualitative Research Methods. 6. Koester, Merrie

Keywords: Educational Semiotics, S.T.E.A.M. education.

Targeted Audience: Pre-service science teacher educators –instructors of qualitative and/or arts-based research methods –classroom science teachers – educational policy makers - S.T.E.A. M. PD facilitators

Cover art and interior photography by Merrie Koester

WHAT OTHERS ARE SAYING...

Merrie Koester has created a well-researched and documented, persuasive and liberating, serious yet playful book that should be required reading for all science teachers and teachers in preparation. Even the structure of this book reflects its message. Its bones are the solidly researched citations, while at the same time it is poetic, artistic prose whose warm empathic stories clearly illustrate meaning. I spent my career creating programs and providing professional development for teaching science as inquiry, insisting that science and art are inseparable—a message Merrie Koester echoes, "Without art, there is no engineering or invention." Koester persuasively illustrates that art also "connects, communicates, and breaks down barriers". Art opens windows to the world for English language learners, poor readers and others too commonly seen as poor students. Equally important, Koester shows us that teaching science as an aesthetic inquiry can be done in the standards-based, ultra accountability context now engulfing education in America. Teachers can't just script and direct student learning: "Artistic science teaching is far less about telling what is right or wrong than it is about revealing what is significant and meaningful so that students can interact with the content and each other." Actively engaging in science and illustrating meaning through art provides common shared experiences through which students and teachers learn together.

Science Teachers That Draw: The Red is Always There is for all science teachers, elementary and secondary, including those wishing to become teachers, and for the higher education faculty who prepare them.

—**Donald B. Young, Dean, College of Education,
University of University of Hawai'i Mānoa**

Sir Ken Robinson calls for an aesthetic awakening in our schools if we are to engage today's students in thinking and learning. Merrie Koester takes us into science classrooms and demonstrates how transformational educational change is possible by teaching through aesthetic inquiry. As she shows, graphic scientific visualization is not that hard and scary; educators only need to open themselves to the possibility and try it.

> —*Richard Siegesmund, Professor and Head, Art+Design Education, Northern Illinois University, co-editor with Melisa Cahnmann-Taylor of* **Arts-based research in education: Foundations for practice**

Merrie Koester offers science teachers a richly illustrated guide to making science more engaging, more accessible, and truer to the imaginative foundations of scientific inquiry, through the visual arts. Her research-based approach empowers students to notice more, connect better, and express themselves visually as well as verbally. This book will be especially valued by teachers of struggling readers and English-language learners, but it provides a powerful resource for every teacher who values creativity and active learning. Koester's approach, based on her own and other teachers' experience, is centered in the power of visual expression to engage students in learning about the natural world, but more deeply it is about caring. Caring for the world, caring about one another. Her book is rich in poetry as well as drawing and in the revealing expressive work and collaborations of students and teachers caring together.

> —*Jay Lemke, Department of Communication, University of California–San Diego; author of* **Talking Science: Language, Learning,** *and* **Values** *and* **Textual Politics: Discourse and Social Dynamics**

For my beloved sons,

Hank and Karl

Contents

Affirmations		11
	What If?	17
	Taking Heart	20
	Visualize	29
1	Science as Aesthetic Inquiry *Creativity, Deep Learning, and Ways of Being Science Teachers and Learners*	31
2	Starting *Mindset and Method*	51
	Drawing in Science Research	54
	PCK, and The Back of the Napkin Test	58
	Edusemiosis	61
3	"Know"tations and the Nature of Science *Creating Narratives*	63
	Mindfulness	70
	Harold and the Purple Crayon	72

	An Intentional Undertaking Designed to...	74
4	**Nurture** *Performing an Ethic of Care*	75
	When Mia Speaks	76
	Bones Digs for Data	79
	Ms. Maya and Muhammad	90
	Do My Shells Make Me Look Fat?	105
5	**Notice What There Is to Be Noticed** *Inside and Out*	119
	Skybeams	121
	"Know"tations of Biotic and Abiotic factors	129
	Lockdown	142
6	**Create New Vistas** *Showing and Telling*	147
	The Story of Chilli the Mutant Crab and the Bo Jangles Chicken Leg	152
	Voice Wordles	166
	Synthesize	169

7	Science, Philosophy, and Aesthetics *Tangibles and Intangibles*	171
	The Tree of "Know"ledge	172
	Failure to Thrive	181
	Energy Transfer and *Umwelt*	182
8	A Model of Science as Aesthetic Inquiry	189
	Changing Teacher Beliefs and Perspectives	190
	Research, Teacher Practices, and Student Learning	193
	Time for Experiencing and Collaborative Reflection	194
	Drawing Artifacts and Assessment	196
	An Evidence-Based Model	197
	Conclusion	203
	Glossary	207
	Epilogue *Notes for the Qualitative Researcher:* *Arts-Based Educational Research and Practice*	215
	References	219

Deep University Press Scientific Board   232

Guide to Authors   237

Merrie Koester's Biosketch   239

Affirmations

I never set out to become a science teacher, but there were few career options open to biology majors in those twentieth century days. I had foregone medical school in favor of becoming a mother, the most fulfilling occupation I have ever had. I tried bench research. I was miserable. So, by default, I earned my teaching certificate, the training for which left this bio major woefully unprepared for my first gig—teaching physical science, art, and aerobic dance with no lab, no art sink, and absolutely zero dance experience.

The job fit my two young sons' school schedules, so I took it, albeit with trepidation. I had never taught physics or chemistry, and what I knew about aerobic dance I had seen on a friend's Jane Fonda video. The science class was filled with students who had previously failed science. In the sinkless art class were students the guidance counselor had nowhere else to place. The dance class was an add-on PE class. My PE students laughed at me. I laughed at myself. This class turned out to be a blast. We organized a bucket brigade to the hall bathroom for our art classes. My students even turned out some respectable work. The science class, however, was a disaster. I was given a science textbook that was ten years old and told to turn in my lecture notes on Mondays. There was no science teaching or measurement equipment. I had three nineteen-year old freshmen in a class of thirty (mostly boys), one of whom had just been released from his second stint in an alternative school for shooting out the windows of his former school. Nearly all students were reading below grade level. Nothing in my life experience had prepared me for such a challenge. I am not by nature prone to panic, but I sweated ten times more in this science class than I ever did in the PE class.

How could I teach science to students, many of whom were functionally illiterate? Certainly not through lectures; however, I knew no other way to do this job—or did I? I did! I recalled three biology professors at Furman University who had taught their subject specialties by enthusiastically filling their chalk boards with clear explanatory drawings. I had been awestruck by the depth of knowledge which must have been required to pull off such compelling and meaningful teaching performances. At that moment, however (even if my life depended on it), I could not have created an explanatory drawing for, say torque and the center of gravity or momentum vectors. Suddenly, the scary reality of my situation hit me. Without having deep content knowledge of physics myself, I was doomed to fail as a so-called science *educator*.

As a *teacher* of physics, I was like a poor musician who barely knew her scales. My students would learn no physics *music* from me until I learned how to put the notes together myself. As it stood, I remembered almost nothing from the two lecture-based physics courses I had taken in college. I had skipped high school physics to take classes I was sure I would enjoy more. After supper on the Friday of my worst ever week of teaching science, I took myself and my boys to the local library and checked out every book I could find that explained physics principles to *kids* using pictures. That night, I practiced drawing funny little cars, pulleys, and inclined planes with little cartoon people doing *physics things*. Over the weekend, I composed a *Phew Phunny Physics* poems. In the process of making this teaching art, concepts I had previously not understood myself became much clearer to me. I could not wait to try my teaching experiment the following Monday. To my delight, my students were immediately engaged, and they started filling their own notebooks with science-themed drawing artifacts. Within a week, we were using physics concepts like Newton's First Law of Motion to explain the Couch Potato Phenomenon. Encouraged by the success of my first ever teaching breakthrough, I made an appointment to see my principal. "May I please teach my science class through art?" I asked, even though I was already teaching this way. His terse reply: "That's ridiculous. Stick to the curriculum." Ah. Accordingly, I sent him his lectures and continued teaching through the arts. However, I lectured whenever the principal visited my class. The solution worked for all of us. Thank goodness, this was only a one-year position, as I am not at all sure I would have been able to sustain my rogue performance.

Never again did I find myself working for administrators who did not recognize the value in teaching science through the arts. At the University of Hawaii, I was enthusiastically encouraged to research and develop arts-based science curriculum as part of my master's degree. For the next twenty five years, I field tested my pedagogy at the middle school level. In the arena of professional development, I found champions in people like Paula Keener, education coordinator for NOAA's Office of Ocean Exploration and Research, who convinced me to publish a series of science education novels. Many professional education organizations provided platforms for me to share methodologies for arts-based science inquiry. I am especially grateful to the support I have received from the South Carolina Department of Education and the South Carolina Science Council. A few years ago, Meta Van Sickle, chair of the teacher education program at the College of Charleston, afforded me a teaching/learning experience that transformed my world. She hired me to teach a pre-service science methods class, which I thoroughly enjoyed. I began giving serious consideration to pursuing my doctorate, an effort I had postponed to raise my sons, now grown young men, who said, "Go for it, Mombo!" I might never have done so, however, had Meta not invited me to participate with the rest of the College of Education faculty in the *The Algebra Project*. Here was a highly organized effort to create mathematics symbology which was participatory, personal, culturally congruent, and most of all highly communicative and meaningful. This was education at its best. I felt I had come home at last! On the last day of class, I sought out Staffas Brassard, one of *The Algebra Project* professors. I told him that that this experience had caused me to understand that teaching was at its heart a process of effecting rich and lasting *signification*. I now understood that the creative arts—the drawings, stories and poems, the creative drama—were all *enriching* this process of signification by *bridging* the strange (the science) with the familiar (the arts). I then shared with him that I was thinking about earning my Ph.D. Staffas smiled and nodded. "You're doing semiotics," he announced. "I'm doing what?" I replied. "Semiotics is how you should frame your doctoral research," he continued. "Here, I'll write the word in your sketchbook." And that was that. The lone word beckoned me. It was both sign *and* symbol of new possibilities.

I eagerly and immediately plunged into the literature on semiotics, soaking up Saussure, Peirce, Kull, Deely, and Sebeok. I quickly surmised that semiotics was much more than science. It was Philosophy. It was Art. It was Teaching. It was Learning. The sheer elegance of Peirce's Theory of Signs stunned my synapses. I could not imagine why semiotics was not included as essential curriculum in all teacher education programs and vowed it would become an integral component of my own science methods classes. In Francois Tochon of the University of Wisconsin, I found not only friend and colleague, but semiotics guru. In Christine Lotter, Bert Ely, Zach Kelehear, Karen Heid, Rhonda Jeffries, and Allison Anders of the University of South Carolina and Meta Van Sickle of the College of Charleston, I found the support to launch my learning onto a trajectory which has propelled me into orbits of discovery I had never thought possible. With financial backing by the USC Center for Science Education, directed by Bert Ely, at long last, after twenty-five years as a science educator, I earned my doctorate, launched *Project Draw for Science*, and now have written this book, a testimony to the significance of aesthetics in the enactment of semiotics in the science classroom. I salute all who have encouraged me along this incredible journey.

What If?

The bell rang, and all of us were both surprised and dismayed. One eleven-year-old boy exclaimed, "but, Mr. Marsh, I don't want to leave!" This outcry was all the more astonishing, given that students were being dismissed for their spring break. For the tenth time, I checked the monitor on my tape recorder to determine that I had not accidentally shut it off during the lesson. I quickly reviewed the pictures I had taken of one of the most artistic teaching experiences I had ever witnessed, regardless of subject matter and grade level. My task as an education researcher was now to somehow re-present the particular qualities I had observed during the unfolding of a multisensory, imaginative science teaching and learning performance which had escaped the atmosphere of mere craft and positioned itself in that rarefied space of *artistic* pedagogical achievement. In this educational performance, teachers and students took turns playing "the lead", each catalyzing and renewing the learning reaction with a new input of imagination and/or insight. To our collective delight, Mr. Marsh's goofy cartoon drawing of a "mutant" crab named "Chilli" had just motivated twenty-five sixth graders to design, sketch, and animate a strategy for helping this hungry ten-legged scavenger retrieve a half-eaten drumstick from the bottom of a trash can on the beach. The entire time, students and teacher were conversing in science, fluently using terms like effort, load, and fulcrum in the most naturalistic of ways. One by one, the students said their goodbyes. I scribbled in my field journal, watched and listened. Many hugs with the teacher, Mr. Marsh, were exchanged. He himself was beaming.

What was happening here? My interpretation: A wide array of science, technology, engineering, and math (S.T.E.M.) standards had just been actively and aesthetically explored in a ways that were participatory, transformative, embodied, present, active, imaginative, caring, empathic, and joyful. The list of credits for this performance rolled in my imagination—one science teacher, age forty-five, with nineteen years of teaching experience, a classroom of students, age approximately twelve, who wanted to grow up to be just like him.

Mr. Marsh had, for the better part of a year, joyfully participated with me and five other teachers in a collaborative action research project exploring the possibilities of teaching science through drawing, especially for students who struggled with reading. Significantly, he was dyslexic himself, having felt the sting of marginalization his entire life. And yet, there were "certain teachers" who stood apart from the rest:

> I had a learning disability as a child, but
>
> certain teachers could
>
> "catch me".
>
> This is who I want to be.
>
> I seek to
>
> change
>
> negative experiences and
>
> inspire
>
> positive views.

Mr. Marsh had clearly become one of those "certain teachers". Phillips and Siegesmund (2013) have noted that "by seeking to engage our students in transformational curriculum, we transform ourselves" (p. 221). Tochon (2013) asked the question, "What do classroom performances communicate?" (p. 25). In his monograph, *Signs and Symbols in Education: Educational Semiotics* (2013), Tochon reminded educational researchers that the process of meaning-making, or *semiosis*, is a dynamic, fluid, situational, dialogic, and contextual phenomenon, not reducible to static method or describable within a fixed lexicon. He challenged educators (myself included) to engage in forms of curriculum inquiry which stimulate *metasemiosis*, "generating meaning-making on meaning-making" (p. 28). He explained that the form communication symbols take and the way they are presented (performed) is the key to whether the educational system functions as intended and deep learning occurs.

In Mr. Marsh's innovative lesson, an imagined, chicken-loving decapod set into motion problem-based learning wherein both science and engineering practices took center stage in *artistic* teaching and learning performances. The symbology he employed to communicate meaning in science was performative, fluid, expressive, held relevance for his students, and actively connected them to the content. The jointly performed acts of drawing connected Mr. Marsh to his students and them to the content, whose meaning he strove to communicate. The take-home lesson: Art made science *matter*. Once that switch was flipped, science knowledge could be applied and bridges engineered to destinations constructed in his students' imaginations. All Marsh had to do was get out of the way. The same kind of transformations were happening in the classes of the other teachers in the research practice. Why? Because art *does* things to people—even convincing them to change their beliefs about science teaching and learning.

TAKING HEART

History has not always regarded science and art as the polar opposites they now seem to be. Before the 17th century in the western world, science and art were regarded as a "dynamic duo" of sorts, each necessary for the fullest understanding of our world. Both scientists and artists take their creative inspiration from nature and proceed through cycles of observation, revision, and testing of ideas. Both science and art involve high level cognitive thinking like analysis, synthesis, and decision making. Long ago, art especially was conceived to encompass a much wider range of experiences than it does today. Renaissance writers, taking their inspiration from the ancient Greek wisdom of Aristotle and others, referred to art as "any act of making that involves intellectual judgment" (Taylor, 1964, p. 46). Along these lines, architects, physicians, rhetoricians, and skillful storytellers all engage in "acts of making" which are art.

In this book, art has been conceptualized as the medium through which human beings develop their intuitive and creative potential, while science is portrayed as a precise, objective, quantifiable way of knowing the world. All arguments have been grounded in my fundamental belief that there exists a natural *mutualistic* relationship between art and science, each with the potential to nourish and enhance the other within the ecology of the laboratory and the classroom. Like complementary colors, they bring out the best in each other. Take a walk on most any university campus today, however, and you'll find the schools of Arts and Sciences are rarely near one another. They have, intentionally and structurally, been created as different worlds. I believe this observation is significant and at the root of many present problems in our educational system today. However, in schools where the arts are being infused into the core curriculum, academic turnarounds for previously struggling students are *happening* (Seashore, K. Anderson, A. & Riedel, E. 2003; Rabkin & Redmond, 2004; Dwyer, 2011). I shall, in this book, report similar findings. How and why have these transformations occurred? What is at the root of these success stories?

According to Siegesmund (2010), a young graduate student named Alexander Baumgarten first coined the term *aesthetics* as part of his master's thesis in 1735 by hybridizing four Greek words related to sensory perception: *aesthesis, aisthanesthai, aisthetos,* and *aisthetikos* (Welsch, 1995; Heid, 2005; Siegesmund, 2010). Baumgarten's aesthetics was about the ways in which interpersonal relationships with self, other, and the world were established through the senses. How many hundreds of thousands of American students have been anesthetized *in vivo* in the science classroom through the didactic, linear, auditory knowledge delivery system traditionally and historically employed?

Siegesmund (2010) elegantly explicated the arguments by eighteenth century German philosopher Frederich Schiller against an educational approach in which intellect alone was privileged. Schiller insisted that "sense and intuition must *infuse* intellect [and that] we must *feel* how we are in relationship to objects and persons around us" in order to be fully capable of reason. He saw the fostering of such connections between the senses, feelings, and intellect as falling in the province of inter-relational *aesthetics*. Schiller went so far as to argue that an educational system which devalues aesthetics "could only produce students who are incapable of reason (Siegesmund, 2010, pp. 84-85). Even though creativity is what fuels innovation in every domain of science practice, we live in a world where the *qualitative, artistic,* and *aesthetic* dimensions of the Nature of Science (NOS) are rarely acknowledged, much less celebrated, in *school* science. Consider the possibilities if this mindset were altered! The wide variance of artistic processes and products might instead be viewed as sources of educative goodness *and* fitness, just as diversity in a gene pool improves the survival chances of a population. Alas, the *No Child Left Behind* movement has been *no* friend to those who would espouse an arts-based, emergent approach to teaching and learning (especially in science and mathematics). Sadly, true and measurable *learning* is at this time in our nation's history is deemed to have occurred only if students can score well on standardized tests, which, by design, favor test-takers from the dominant culture who are fluent in English and the idioms and experiences of that culture. Teaching to these tests through "drill and kill" has become standard practice, especially in science education.

I am an apologist for teaching science in ways that create time and space for work and thinking that is qualitative, emergent and creative, not just numbers-based, procedure-driven, and geared toward mastery of knowledge that will be on a test. I am not calling for some form of religious conversion; nor am I suggesting that art smocks replace test tubes. I certainly do not seek to displace in any way the important traditions of situated, constructivist scientific inquiry (Brown, Collins, & Duguid, 1989). I seek, instead, to convincingly argue that intrinsically *inexact*, artistic, and poetic ways of understanding and communicating the meaning, context, and processes of science phenomena can catalyze learning reactions resulting in a broad, meaningful spectrum of understanding for all students--not just dominant culture students who can read well.

Through my research and my experience-based apologetics, I have attempted first to identify those factors which may stand in the way of teaching science as aesthetic inquiry, as the effort can require significant conceptual change. Throughout my long career as a science educator, I have experimented with teaching science through *many* of the creative arts. This particular book is about what happens when *drawing* is deployed as the main signifier of intended meaning of science content as well as a vehicle of collaboration, communication and aesthetically charged discovery. The whole idea is to make science accessible to all and to return the senses to science sense-making.

Five hundred years ago, polymath Leonardo da Vinci wrote in his notebooks that the artist's skill in visual perception through the senses was essential to the scientist who sought to unravel the mysteries of the natural world. He perceived that both science and art were "expressions of human potentialities" (Wallace, 1966, p. 117). Da Vinci compared the practice of disconnecting the senses from scientific learning to a "sailor who gets into a ship without a rudder or compass and who can never be sure where he is going" (MacCurdy, 1941, p. 910). Scholars have remarked on this quote for centuries, but truly I believe that what Leonardo was trying to say was that a ship without a rudder will hit stuff. Similarly, we run aground in our science teaching efforts when we communicate to our students the distorted idea that science is only about memorizing facts, plugging numbers into complex formulae, and following a specific method to arrive at pre-ordained conclusions, printed in a book. Teaching science in this manner has the added problem of marginalizing students for whom both the language of English and science are foreign. Art can open wide channels for deep learning of science for all students, even if they can't read.

Trefil (2008) has argued for a science education stance which is more inclusive and whose goal is to effect a type of science literacy which will allow a human being to "deal with issues that come across our horizon, in the news or elsewhere" (p. 28). This teaching strategy of making learning relevant is not a new one. In 1877, Charles Sanders Peirce published an article in *Popular Science* called "How to Make Our Ideas Clear". In this essay, Peirce argued that a full understanding of any concept required not only being able to come up with a definition for it, but also connecting it in some way to one's daily life and lived experience. Educational philosopher, John Dewey (1938), who worked with Peirce at the University of Chicago, called such relevant experiences both *educative* and *democratic*. Today, contemporary scholars in the field of culturally relevant pedagogy (Delpit, 1988; Ladson-Billings, 1994; Lee, 2003) declare that we need to find ways to generate congruence for culturally and linguistically diverse students between the mostly foreign culture of academic science and their own natural ways of knowing the world. I hope that my own work can add to these conversations.

I have long argued that an *educative* science symbol system should be amended to include a non-linguistic, fluid, aesthetic syntax which is personal, relevant and imaginative, employing symbols which may be mimetic, expressive, or iconic, and communicated (performed) through verbal, visual, and/or kinesthetic means. My theoretical framework is congruent with Lemke's (2004) statement that "scientific communication and scientific literacy are fundamentally multimodal" (p. 1). The narrative which follows documents the ways in which science teacher researchers used drawing as their dominant building material to construct "semiotic spaces" in which previously failing students acquired aesthetic capital and agency, which most brokered into improved academic achievement. Both teachers and students underwent significant forms of transformation, made possible, I believe because of the aesthetic nature of this curriculum inquiry process. Our arts-based educational research was performed on the stage of participatory (or collaborative) action research, a genre of qualitative inquiry which Kemmis and McTaggart (2005) have characterized as having the following key cyclical features: "1) planning a change; 2) acting and observing the process and consequence of change; 3) reflecting on the processes and consequences; 4) re-planning; 5) acting and observing again; 6) reflecting again, and so on." (p. 563). While progressing through my capacity building class, *Teaching Science through Drawing*, each teacher on this participatory action research team *experimented* with improvisatory, *artistic* ways of being a science educator. Each created his or her own innovations for teaching science through drawing to meet the needs of their particular students so as to widen as much as possible the semiotic field. What a privilege it was for me to sit at their feet and simply marvel at their artistic performances as they, too, became Science

Teachers Who Drew! They, like all of us, are natural born artists. The potential had been there all along. They just didn't know it.

    I view science literacy as a pluripotent affordance without which we can become marginalized or even manipulated, given that decisions based on science and technology are made on our behalf every single day. When achieving science literacy depends primarily on English reading and listening fluency, however, there can be profound problems. The robust correlation between reading comprehension and science achievement test scores has been well documented. (Carnine & Carnine, 2004; Cromley, 2009; Hapgood & Palincsar, 2006; Lee, Fradd, & Sutman, 1995; Norris & Phillips, 2003). The following NAEP test scores in reading and science for 8th graders across all states for 2011, which is similar to all previous years' results, make visible the true nature of the struggle:

NAEP 2011 Science and Reading Assessment Summaries

Student Category	% Scoring Below Basic Level in READING	% Scoring Below Basic Level in SCIENCE
All students	25	36
Low income Families	37	52
English Language Learners	71	83

The poorest readers and lowest achieving science students in America often come from either low income families or from culturally and linguistically diverse families. These students, far more so than dominant culture students, are often tracked into the lowest level classes with the least highly qualified teachers and with consistently low expectations made of them (Oakes, 2005).

Thankfully, marginalized students have had many powerful champions. In the 1980's, Robert Moses created a radical new approach to teaching mathematics called *The Algebra Project* as a way of confronting the epidemic of math and science illiteracy among Blacks and minorities (Moses & Cobb, 2001, p. 11). Moses believed that for too many young people, mathematics was a game of signs they were unable to play…a manipulation of a collection of mysterious symbols' (p. 122). If you give a person a map but he or she doesn't understand the symbols used in the legend, the map is worthless as a way-finding instrument. *The Algebra Project* is all about taking *meaningful* trips to life-altering destinations:

In the spring of 2011, I and some twenty education faculty at the College of Charleston found ourselves on a chartered bus, at the front of which was an ebullient gentleman who explained that this would be a trip like no other we had ever taken. He was clearly a gifted storyteller, his voice resonant and compelling, his thick, dark eyebrows spiking with each exclamation. I fidgeted in my seat. Something phenomenal was happening, and we hadn't even left the curbside. Over the next several hours, the bus stopped at destinations, which defined the history of a Charleston I had never been told, even though I grew up here. That day, we traveled through a trail of oppression and resistance and hope from churches and waterfronts to ironworks studios and cemeteries. As has always been my habit, I had my sketchbook with me. I sketched out a line drawing in the shape of Charleston's peninsula, drawing in those streets along which we traveled, placing symbolic icons to represent the meaning each stop had held for me. Drawing has always been my go-to way of learning and processing information. As I drew, words and images became linked with story and place in a symbolic interactional experience I would never forget. The structure of my brain was changed forever, and I knew it. I could hardly contain myself. It was as if someone had slipped me a dose of Doctor Buzzard's Hoodoo Water, a bottle of which I had just seen on display at the Avery Research Center, the last stop on our magical mystery tour. This is what happens when you take an *Algebra Project* trip! Key ingredients of this powerfully aesthetic and socially constructed curriculum include "cooperation and participation in group activities, as well as personal responsibility for individual work" (Moses & Cobb, 2001, p. 121.)

Like my own approach to teaching science through drawing, *The Algebra Project* curriculum invites students to draw pictures to "portray particular aspects of the experience" in *any* way they find meaningful. Another close parallel to my methodology was *The Algebra Project's* emphasis on symbolic representation and collaborative meaning-making. Moses and Cobb's (2001) book, *Radical Equations*, together with the training by *The Algebra Project* facilitators gave me the impetus and inspiration I had been seeking to, at long last, develop my own doctoral level research on drawing as essential symbology for meaning-making in the science classroom. As a result of this research, I have now documented that teaching science through drawing using an aesthetic paradigm "works", even by traditional standards of measuring "success". Student test scores improved, with the biggest gains by previously struggling students. Little by little, there are forming all over the United States, tribes of other science educators who are experimenting with arts-infused S.T.E.M. curricula because they dared to answer the question, "What if?" Something very exciting is happening here. Won't you join us?

Visualize…

1 SCIENCE AS AESTHETIC INQUIRY

Creativity, Deep Learning, and Ways of Being Science Teachers and Learners

Ralph Ellison (1963) wrote that one of the things missing from the discussion of American education is imagination. Rolling (2013) has argued that schools, by design, *under-develop* creativity in favor of the "development of citizens who are easy to categorize, easy to sort into cubicles or assembly lines, and easy to manage" (p. 42). In their text, *Who Killed Creativity...and How Do We Get it Back?*, Grant and Grant (2012) recounted the results of research on K-12 student self-identification as "being a good artist". In kindergarten, *all* students considered themselves as being artistic and creative. By the eleventh grade, only one student raised their hand when asked "Who thinks they're creative?" Along these same lines, Trefil (2008) described in his landmark text, *Science Matters*, a phenomenon which he called 'the Great Turnoff' to science, a disheartening event in the lives of many adolescents that occurs sometime around middle school:

> Normal curiosity about the world seems to turn into disdain for, and perhaps even a fear of, things scientific...In eighth grade, a student is likely to encounter more vocabulary words in his or her science class than in English...What should have been a vital, engaging hands-on subject was turned into a dry exercise in rote memorization...The high school years in

America are a continuation of the turning away from science that starts earlier (pp. 131, 137, 145).

Trefil recognized that when and where there is no creativity being fostered, students cannot appreciate the true nature of science. He claimed that because of the way science is traditionally being taught, we have become a nation of science illiterate adults, with most citizens not having enough knowledge about the "physical universe to deal with issues that come across our horizon, in the news or elsewhere" (p. 148). How can we get this vital subject to *matter* to our students, especially those who are poor readers and who view science as being too hard or inaccessible to them? This question is what has driven my research. I believe an answer lies in teaching science as *aesthetic* inquiry, as the purposeful attempt to create learning *experiences* which are at once beautiful and moving.

In framing my research, I was deeply influenced by the work of Maxine Greene (2001), who defined aesthetic education as "an intentional undertaking designed to nurture appreciative, reflective, cultural, participatory engagements with the arts by enabling learners to notice what there is to be noticed" (p. 6). According to Greene, when we teach using an aesthetic approach, we create new connections in experience, allow new patterns to be perceived and open "new vistas" of understanding. Greene's sense of aesthetics also guided the coding of my data, providing me with a critical "noticing" checklist. For example, what types of *nurturing* were going on in the classrooms of teacher participants? What other *arts* were being employed as part of teaching innovations? In what ways were teaching and learning *participatory* and *reflective*? In what ways were teachers enacting *culturally inclusive* practices? What forms did *engagement* take? What emerged from my data analysis and reflections with my team was a tentative model of science as aesthetic inquiry. It's still very much a work in progress.

By their very nature, aesthetic experiences have the power to transform (Bresler, 2006; Dewey, 1934, 1938; Eisner, 1972; Siegesmund, 2010; Uhrmacher, 2010). Uhrmacher (2010) explained the differences between an aesthetically oriented curriculum and *other* models, which "do not account for the life-enhancing moments that do not rely on measurable objectives" (p. 193). He proposed an "Aesthetic-Transformative" model of curriculum implementation whose purpose was for students to "attain aesthetic capital", effectively adding to Pierre Bourdieu's (2003) triumvirate of economic, cultural, and social capital and Coleman's (2003) construct of "human capital", which is acquired when one learns a new skill or gains new knowledge (p. 194). He cited Adjibolosoo (1995), who explained that "aesthetic or artistic capital brightens or enriches peoples' lives" (p. 181). According to Uhrmacher, aesthetic capital may cause a student to feel or act differently and to undergo a radically different way of being in the world. Significantly, the Aesthetic Transformative approach is an emergent one, wherein "methods can vary widely" and in which the outcomes are "related to its purpose" (p. 195).

Over five hundred years ago, Leonardo da Vinci recognized that not only was *vision* the dominant human sense but also that *drawing* was an extension of that sense. Deep study of da Vinci's work reveals that here was a mind that recognized that to draw something is to know it *and* to feel it, the hand providing conduit from perception to paper. Through the aesthetic, sensory experience of drawing, profound understanding can be achieved. Johann Heinrich Pestalozzi (1894/1973), who devoted his career to the education of impoverished children, believed that children could come to an interpersonal relationship with themselves and the world through the dedicated practice of drawing:

> The general advantages resulting from the early practice of drawing are evident to everyone. Those who are familiar with art are known to look upon almost every object with eyes different as it were from a common observer. One

who is in the habit of examining the structure of plants and conversant with a system of botany will discover a number of distinguishing characteristics of a flower, for instance, which remain wholly unnoticed by one unacquainted with that science. It is from the same reason that even in common life a person who is in the habit of drawing, especially from Nature, will easily perceive many circumstances which are commonly overlooked, and form a much more correct impression even of such objects as he does not stop to examine minutely, than one who has never been taught to look upon what he sees with an intention to reproduce a likeness of it (Pestalozzi letter of 1819, in Anderson, 1931, p. 179).

In my master's research at UH, I studied ways in which a wide array of creative arts (music, dance, drawing, painting, creative drama, writing, photography, etc.) could serve as media for meaning making in science. In my thesis, I wrote that poets in particular are capable of taking what seem to be mundane objects and putting them square into our imaginations, making visible their profound realizations about *science* matters. The best poems arise through the practice of intense *noticing*. Consider these two American haiku from the book, *Borrowed Water* (Los Altos Writers' Roundtable, 1966):

O stuporous snail

    What night time frolics took you

        On this tinseled trail?

            ~ Tashjian

Under the dark earth

The majestic oak is bursting

Its acorn prison.

~ Rutherford

In less than 75 characters (far less that the maximum allowed for a Tweet) each of these haiku communicates the possibilities of *becoming,* of *transformation.* They cause us to look at snails and acorns in completely different ways. How many of our students sit stuporous and dazed before us, when their real selves are poised to frolic and burst forth from the imprisonment they feel while in school? What might we do differently with the curriculum so that our students won't feel this way? The standards only tell us what to teach, not *how.* Can taking an artistic orientation to teaching and inquiry liberate students from the prison so many perceive school learning to be? During my doctoral research, I challenged teachers to play with the curriculum and to see what happened if they frolicked a bit—and let their inner snails soar. Teaching science as aesthetic inquiry can leave trails of tinseled delight, and transform attitudes from stuporous to stupendous and acorn-level knowledge into mature oak-level understanding. It's not as hard or scary as you may think to be this kind of teacher. Will you try it?

How might the common snail or an oak help us learn core S.T.E.M. concepts? I recently learned that snail slime is being marketed as a beauty product! There's a science and engineering learning hook if ever there was one. The oak may look like any other tree, but it's played a mighty role in global industries and even politics. The Emancipation Oak has become a symbol of all that is possible if one is resilient. How might students be compared to acorns? From what prisons have humans escaped in order to live strong and free, like the mature oak? How are pines like the trailblazers among us? Pines appear first in the process of forest succession, and then stimulate more life in the midst of their own death by fire. Yes, all this science can arise from discussion kindled by two tiny, but powerfully evocative poems. The arts move us.

There are infinitely many, meaningful stories in science just waiting to be told, written about, and illustrated by you and your students. They have always been there and will include even struggling readers into the wonder that is science. However, performing such pedagogy is no easy task. There are arguably few bodies of extant knowledge more difficult to convey than that which science teachers are expected to meaningfully communicate in the hopes of effecting science both literacy and appreciation. *Science for All Americans (1989),* the official publication of Project 2061 by the American Association for the Advancement of Science, and most recently, the *Next Generation Science Standards* (2013) outline specific recommendations for learning in the following areas of science, technology, mathematics, and technology with the stated goal of a science literate citizenry: The Nature of Science, The Nature of Mathematics, The Physical Setting, The Living Environment, The Human Organism, Human Society, The Designed World, The Mathematical World, Historical Perspectives, Common Themes, and Habits of Mind. That's a ton of content! How does the science teacher, especially the novice, go about the task of deconstructing these intimidating marching orders with a growing population of students who are scoring below basic proficiency levels in both reading and science?

Van Sickle and Spector's (1996) study on *caring* science teachers determined that taking the time and effort to deepen one's subject content knowledge was a consistently observed characteristic. Many have argued that an *aesthetic* education, by its very nature, *engenders* an ethic of care ((Pestalozzi, 1894/1973; Dewey, 1934; Greene, 1971; Nussbaum, 1997; Noddings , 2005; & Siegesmund, 2013). Pestalozzi recognized that without an ethic of care in the classroom, the child could not experience "wholeness".

For this state of being to occur, Pestalozzi proclaimed that children's learning must take place in an emotionally safe environment, one in which they felt cared for and were allowed to think for themselves (Siegesmund, 2010, p. 85). Significantly, during our study, students who had heretofore not been motivated to learn science at all were now enthusiastically engaged. They were honored by their teachers' research efforts on their behalf. Finally, they could *show* what they could do, rather than stumble in attempts to *tell*. Van Sickle and Spector (1996) have determined that, in addition to taking the time to deepen their own content knowledge, caring science teachers established personal relationships with their students and helped them to understand that all living things are interdependent. Students were nurtured and nurtured each other. Significantly, the teachers in their study had students who consistently performed at high levels of achievement. Every teacher in my study also reported improved test scores by previously struggling or failing students. This ethic of care will be more fully explored in Chapter 4.

Through aesthetic inquiry, art and science combine in powerful, lasting learning reactions. For example, the student who is asked to create a drawing, choreograph a sequence of movements, develop a poem, story, or role play about a science phenomenon is called upon to capture the essence of multisensory encounters with his or her world in highly concentrated, descriptive, *aesthetic* language. To accomplish any of these *artistic* tasks, the student must engage in observation, analysis, decision-making, and synthesis – exactly the same cognitive skill set required of the *scientist*. Drawings, poems and stories can, if invited, burst forth from the heart of the person who suddenly feels them and "knows" something in ways that are made both personal and *real*. Now, the "something" matters to them. They sit up and pay attention. They even start making better grades. Their parents write thank-you letters. Teachers in the study reported all of these positive outcomes. In each case, the implantation of aesthetics into the curriculum disrupted and corrected previous states of educational arrest. When a teacher creates opportunities for sustained noticing of the natural world or the solving of S.T.E.M. problems that might touch students' own lives, deep learning happens. Students thirst for such learning. They also want to know what touches you, the teacher. When you yourself experiment with being an artist, you become more human to them.

In lieu of a lecture about barrier islands and coastal ecosystems, I once wrote the following poem for my students:

Field Trip

All across the island,

creatures are animating,

claw-to-claw, hand-in-hand,

beak-to-beak, mucus to mud.

My son pulls down his cap, while

I adjust my sarong,

the one with the fish on it, and

we set out across the

well-worn path

from the beach to the marsh.

I remember last year,

when he was much smaller,

we only made it over the dunes

and through the slough,

where the mosquitoes

ambushed us.

This year, more prepared,

we plunged into the

forest, patted the oaks, and

crushed berries between

our fingers, while avoiding

raccoon scat.

Approaching the marsh at last,

I smiled, and my son

began to wave

to fiddler crabs, who were most assuredly

celebrating our arrival.

Through the medium of poetry, my students came to know me as a mother, a nature watcher, and part of a cycle of life, out there with my son and the clawed and beaked things. Through poetry, I invited them to visualize taking just such a field trip themselves, which, of course, we did, armed with our trusty sketchbooks. We talked about what kinds of things might be happening when living things "animated", including the fascinating topic of generating scat of all kinds and the forensics of footprints. They acquired deep learning of life science standards. Many times in the last twenty-five years, I have run into previous students (now in their twenties and thirties) who say they still have their middle school science sketchbooks. They still remember dancing to the "Science Rap", writing "Really Rotten Rhymes", choreographing "The Tectonic Drag", and producing "Rock Videos". In many cases, I created joint lessons with language arts, music, media and visual arts faculty. Some school districts now employ arts integration specialists. If you yourself don't feel "gifted" in an arts area, then casting yourself in the role of arts *student* will communicate to your students the powerful lesson that we are all *learners*. The goal is for students not only to remember but *appreciate* the science because of the connections aesthetically afforded them through the arts. The science through the arts products students create will amaze you. Like all art, student work should be shared. How you do so in this digitally connected world is open to endless possibilities.

I believe the new S.T.E.A.M. (or S.T.E.M. and the Arts) movement will inspire teams of teachers to work together to effect arts integration in unprecedented and important ways and that all will benefit from such educational reform. Even in an arts-infused curriculum, however, semiotic pathways can be cut off for ELL students if instructions and materials are provided in English only. Participating teachers in our action research paired new ELL students (mostly Latinas and Latinos) with fluent bilingual seat mates. The bilingual students helped their partners create science drawings labeled in both Spanish and English. The ELL students, in turn, taught their classmates the Spanish terms for science vocabulary. Teachers reported very positive learning experiences for their ELL students. In every way, we have discovered that teaching science through drawing immediately makes signification, and thus meaning-making, possible for students who are not fluent in English. Such students, previously marginalized by their lack of English language skills, now felt included and nurtured, and told us as much.

My research has demonstrated that an aesthetic approach to teaching science through drawing empowers students with significant agency. I have come to believe that students who are denied the opportunity of exploring science through aesthetic, sensory pathways may manifest their frustration through symptoms of resistance, defiance and dis-ease caused by feelings of being dis-connected, de-valued, and even de-moralized. I invite you to listen to Crystal, a special education teacher, who (in spite of the fact she was not certified to teach science) found herself teaching this much dreaded subject in a self-contained classroom of special needs middle school students:

> I attended school in a very rural farming community.
> One teacher taught the first eight grades.
> This meant that we students had to be
> very self directed and collaborative
> with our classmates.
>
> In my early grades, science was
> not really a subject by itself.
> The *sciency* part of history meant
> We made parfleche—raw hide bags
> like the first peoples
> who lived here.
> We made sod and our own sod houses.
> We pondered why the Dust Bowl hit, and

how cholera traveled,
killing
Native Americans and whites alike.
We learned how the Sioux were
given blankets infected with disease

In fifth grade, that changed.
Now we read about experiments,
but we did not do them.
My friends and I spent hours
outside at the river, conducting
our own experiments.
But we did not know we were doing science.

I thought science was something you
did at school, and not in the rest of the world.

Books, ideas, and more ideas became
my delight.
They kept me sane during the long
Aloneness I have gone through.

I am attracted to vast, desolate places.
What matters to me is the Milky Way—
its thick blanket hugging the prairie sky,
whispering that the world is truly
full of awe.

The Gobi Desert matters to me—
its sky in a slowly changing dance.
In Chaco Canyon, Mesa Verde, the Grand Canyon,
if you listen,
you can hear the whispers of
those who came before.

What matters to me is being able to see

ten miles into the distance,
to know the smell of a rainstorm coming,
to see the clouds build and swell and know
with certainty
how much violence they will hold.

I have only to stop and breathe and
pay attention.
Nature returns her reassurances.

Once, a bobcat chased a cat onto my porch.
Then, it stopped to stare me down
Through the glass paned kitchen door.
A blue heron, my father's favorite,
flew alongside the car,
a winged escort to the cemetery.

And once, I got to swim in
a kelp forest.
My high school had fifty students enrolled
across all four grade levels.
Again, science came out of a book.
The one time we got to look in a microscope,
I pretended I could see
what everyone else was seeing, but really,
all I saw was my eyelash.
In college, science was my least favorite class.

I took the ACT and earned a scholarship to university.
I did not know my highest score was in science.
I was too busy dropping
Freshman Chemistry.
Unlike my classmates, for whom this was a
review,
I did not know any of this equipment.
My results were never close to what

they were supposed to be.
I was afraid I would lose
my scholarship.

I still remember my T.A.'s name—Ita.
Though I am sure he was very knowledgeable,
his English was not clear enough for me
to understand him.
When we asked him to clarify an answer,
Ita would repeat what he had just said,
only a little louder.
If we still didn't understand, the repeating
would escalate into near shouting.
I realize now he was likely as
frustrated as we were.
Poor Ita.

After that, I discovered
Geology.
It was more real to me.
I loved it.
Later, after college,
I decided to take a
Biology class.
But once the labs started,
I was lost again.

I earned a master's degree in
Elementary and special education.
The science methods class once again
taught out of a textbook.
I became highly qualified to teach
reading and writing, but did not once believe
I could do so in science.

Then I moved to The South,

where, it seems,
poverty and racism and hopelessness are
the Norm.
Two boys were killed in the front yard
down the street from my house.
I have a concealed weapons license and am armed.
My kids – the ones I teach- don't know the unspoken
rules of the middle class.
They do their best to make sure you
reject them quickly,
so it won't hurt so much.
The longer they wait,
the more it
will hurt.

Together, we are on an island in the school—
a self-contained unit of Special Ed
seventh graders, whom most believe
are expendable.

No way.

I worked extra hard to learn more science,
Taking one workshop after another.
In the meantime, I brought my kids outside into
Nature and its science.

In September, we started drawing everything!
They were getting it—
even passing grade level science tests.

Dare I hope that maybe,
just maybe,
one day,
when they do get a good science teacher,
at least one will escape all this?

Crystal arrived into adulthood and her teaching career believing that *school* science was not remotely educative. To the contrary, her experiences in the science classroom had been oppressive and even demoralizing. I, too, had seen my eyelash during my first encounters with a microscope in middle school. My teacher, Mrs. Wehman, taught me to breathe slowing and relax both eyes. Feeling slightly silly and self-conscious, I followed her instructions. Suddenly, a glorious micro-world burst into view. It had been there the whole time, of course. I adored Mrs. Wehman.

Crystal enrolled in my study because she was intrigued by the possibilities of learning a different way of being a science teacher – one that was more in line with her own artistic ways of being and knowing the world. Although she was teaching out of field, she became the kind of science teacher who inspired our entire research team. She actually had a much easier time of making the leap into an artistic way of teaching science because she had not found success as a learner in the traditional science classroom. However, based on her reflections, she failed to believe she was a "good" science teacher because she did not have a degree in science or an encyclopedic command of science vocabulary about which she could lecture with authority. I was filled with great empathy and admiration for her. She overcame her insecurities about "being bad" at science and managed to inspire her special needs students to perform science learning *concertos*. They had it in them the whole time. It took Crystal to see and to nurture their gifts. Like the best of teachers, she empowered her students to learn by finding wavelengths of meaning-making, which resonated with their "special" and wonderful ways of being *people*.

Lortie (1975) documented that teachers will default to teaching in the ways they have been taught, even if those teaching practices are faulty. Classically "trained" science teachers (myself included) have largely been taught our subject matter content through didactic lecture and explanation of a recognized standard body of knowledge; thus they, too, have become lecturers and explainers. Science teachers' beliefs in the efficacy of transmission modes of knowledge delivery are deeply entrenched (Borko & Putnam,1996). In my twenty years of facilitating professional development for science educators, the idea of drawing for and with their students and guiding their students to imaginatively embody science concepts often has often prompted high levels of concern (at least at first) for non-artist science teacher participants. Similarly, Latta and Baer (2013) have documented that the "terrain" of aesthetic inquiry is one "often feared and deliberately avoided by educators" (p. 94). Many (if not most) science teachers were as students rewarded for their ability to march down the memorization highway, successfully navigating their way toward right answers and the good grades that came with their production. In keeping with this mindset, many science activities marketed as promoting "inquiry" are designed to progress toward the same outcomes every time they are employed by students. Curious students (many labeled ADHD), who seek to diverge from the instructional protocol are quickly brought back into line by those teachers who fear losing control of the learning process. Such "experiments" are no more *science* than a paint by number activity is *art*. As long as this right answer/same outcome mentality is driving science education, the creative nature of science cannot be appreciated by the learner.

Elliot Eisner wrote and presented extensively on the subject of what the arts can do for education: The arts teach us that "everything interacts"; that "nuance matters"; that surprise is natural part of all authentic inquiry; that "slowing down perception is the most promising way to see what is actually there; that "the limits of language are not the limits of cognition"; that the embodied enactment of understanding is an important way of discovering what a student knows; and that open-ended tasks permit the exercise of the imagination (2008, Lowenfeld Lecture, NAEA Conference, New Orleans). Throughout his long career, Eisner placed a high value on multiple ways of representing understanding, frequently citing Michael Polanyi's (1966) observation that we all know more than we can tell. Similarly, University of South Carolina professor Gloria Boutte (1999) has consistently called for ways of increasing a students' cultural *and* science competencies by tuning into *multiple* channels of learning. In her book, *Multicultural Education: Raising Consciousness*, Boutte (1999) shared this quote by Robertta Barba (1995): "Changing the ways that science, math, and related technology courses are taught involves changing the ways that knowledge, teaching, and learning in these disciplines are viewed" (p. 179). Boutte proposed that science should be taught, not as a "single way of investigating the world", but rather as a "multifaceted construct" (p. 180). I strongly believe that teaching science as *aesthetic* inquiry is likely to provide multiple channels of learning and different ways of arriving at "knowing".

The authors of the *National Research Council's (NRC) Framework for K-12 Science Education: Practices, Crosscutting Concepts, and Core Ideas* (2012), have called for research which asks questions about teaching practices which "move students along a path from their initial understanding to the desired outcomes" (p. 30). The authors of the *Next Generation Science Standards* (2013) have encouraged the creation of "programs which allow science teachers to acquire effective strategies to include all students regardless of racial, ethnic, cultural, linguistic, socioeconomic, and gender backgrounds" (Appendix D, p. 17). They fully acknowledge that while they will draw from existing research literature to identify effective classroom strategies that enable students to engage in NGSS, they will also seek new research agenda. I firmly believe that teaching science through drawing and the creative arts is a highly effective way of making science meaningful to the poor reader, for whom achieving science literacy in traditional, linguistically heavy science learning environments can be nearly impossible. To be sure, developing one's artistic ability and deepening content knowledge takes both time and a profound ethic of care. I have only to return to the stories of the teachers in this study to know what transformations can occur when such efforts are made.

My research is not so much about finding one *best* way (because there isn't one), but rather about discovering what might be different if we (my participating teachers and I), were, to borrow Lather's (1993) phraseology, *unjam* prior ways of knowing and being science teachers. In science and in art, epiphanies occur during moments of improvisation. I hope that this narrative will inspire science teachers to adopt a more flexible, artistic, emergent teaching methodology, wherein surprise becomes a welcome guest, and unexpected outcomes are recognized as being fundamental to the nature of both science and of art.

2 STARTING

Mindset and Method

I dedicate this sketchbook to my own amazing self, in celebration of the artist and the scientist within me. By thus inscribing their brand new sketchbooks, which I had given to each of my teacher participants, our year-long journey began! The pseudonyms they chose for themselves were apt metaphors for their personalities, as will be seen. Together, *Marsh, Sky, Eartha, Maya, Crystal, Bones*, and Merrie (my real name) formed a collaborative action research effort we called *Project Draw for Science*. Marsh was the only one of the teachers who had received any formal drawing training, though Bones' father was a graphic artist (so she was very comfortable drawing). Sky was a lifelong "doodler". Eartha and Crystal expressed no overt anxiety about drawing, even though they had no training. Maya, however (the youngest and newest teacher in our group) fretted at the beginning of the study that "she couldn't draw a lick," and expressed a high level of concern about drawing in front of her students. What Maya perceived as a personal shortcoming, however, immediately served to endear her to her students, as she dared to make herself vulnerable by drawing in front of them, as she said, "like a nine-year-old".

Every teacher on the *Project Draw for Science* research team shared the goal of effecting engaging, challenging, standards-based science inquiry, especially for those students who had previously failed to achieve success in the science classroom.

What they had been doing was not working for their struggling students. They eagerly embraced the research question—What will happen when I teach science through drawing, using pedagogy which is aesthetic in nature? Together, we envisioned drawing as mediating sense-based learning and heightened, fully present noticing—practices which were consistent with a practice of aesthetic inquiry (as defined by Greene). We also worked under the assumption that when framed as semiotics, *education* "works" when *signification*, that is, successful communication, works. Wisely (1994), made the following important observation about what makes for effective communication, the keystone of all effective teaching:

> Whether communication is taking place between two people or among many people, [many] elements [determine whether the communication is a success or a failure]: the form and source of the information to be communicated; the symbol system in which the information is encoded; the medium on which it is fixed; the channel which delivers it to the intended receiver; the fields of experience shared by the sender and receiver; the environment in which communication is taking place; and the form of feedback…It is easy to see that a message must be carefully planned, designed, produced, and delivered to increase the probability of successful communication" (p. 92).

In this one synopsis, I believe Wisely has described what situational elements must be present in the moment of the teaching/learning performance if the educative process is to achieve deep learning. At the heart of this communication "ingredients" list, Wisely has placed the "symbol system in which the information is encoded" as being of central importance. His description of the work of the teacher who would effect successful communication—planning, designing, producing, and delivering—are exactly the same as would be employed by an artistic director wishing to pull off a moving stage performance. Science and mathematical symbol systems, (and the cultures of which they are a part), especially because they are canonized and privileged, will remain indecipherable to any for whom translation via shared experiences cannot be established. The inevitable result of failing to communicate science meaning can be summarized in this simple drawing:

The goal of *Project Draw for Science* was to create performances in which communication not only succeeds, but deep learning occurs. To this end, we engaged in progressive, participatory, metasemiotic acts of curriculum inquiry. Our research took place over an entire school year in four site categories: 1) a middle school science classroom; 2) online Skype sessions; 3) in each participant teacher's individual classrooms; and 4) the *Project Draw for Science* wiki discussion board. Over the course of our time together, teacher participants worked to improve their capacity to use drawing to *aesthetically* explore both science content and practices with their middle school students. I challenged them to actively visualize themselves as performing artists, each having the potential and the ability to improve the landscape of science education in the United States.

In the praxis of arts-based methodologies, the "method of discovery *is* the discovery" (Richardson, 1997, p.88). An aesthetic approach to research (and teaching) permits and allows for many individual and different outcomes, more significant for their variance and innovation than any other contributing factor, because they allow the teacher the dignity and autonomy to craft individualized teaching strategies that are congruent with the needs of his or her students. Like independent artisans, teachers pick up new techniques, activities, and materials that fit their own styles and adjust them based on their goals and experience. In short, there is no one best way to *be* an artistic science teacher. Arts integration specialists Freeman, Seashore, and Werner (2003), have explained that capacity building can involve either the teacher learning and practicing the art form themselves (the paradigm we adopted) or with teachers working alongside arts specialists as observers, learners, and helpers. A high degree of jointly produced learning can take place with either approach, especially if arts specialists have also participated in science capacity building training. Imagine the powerful pedagogy which might launched by such twin engine teams of teachers! I could also envision reciprocal arts-science-ELL capacity building colloquiums.

During our capacity building sessions (a course I developed called *Teaching Science through Drawing*) I wanted as much as possible to trigger a creative chain reaction and then get out of the way. However I also stood ready to facilitate active transport when the way in seemed blocked. We would schedule extra Skype sessions for arts skills building, explanation of difficult to teach science concepts, and confidence-building with emergent, arts-based teaching practices. I also facilitated group problem solving exchanges on the research wiki.

DRAWING IN SCIENCE RESEARCH

We began our inquiry by reflecting on the growing body of literature on the subject of using drawing in the science classroom. I shared several important research studies, which had all documented the formative assessment value of this practice. We began with the work of MIT research scientist, Felice Frankel, who conducted an extensive NSF-funded study (2007-2010) on the use of student drawing to identify underlying misconceptions about science. In Frankel's *Picturing to Learn* project, science students from Harvard, MIT, Duke University and Roxbury Community College were asked to "create a freehand drawing to explain to a high school senior" various scientific phenomena. Rubrics for accuracy were developed by the participating university science faculty. The primary findings of the program were that drawings by students from these Ivy League schools revealed a plethora of science misconceptions and confirmed that, even in Ivy League schools, what is supposedly "learned" in science is often not what is intended by the teacher.

Sharon Ainsworth (2006, 2008, 2011) from the University of Nottingham has done extensive research in the area of arts-infused teaching and learning in the science classroom and is taking a leadership role in this area of arts-based educational research. She echoed Howard Gardner in her work and bemoaned the lack of arts methodologies in the science classroom in particular: "In science classes, it is rare that learners are systematically encouraged to create their own visual forms to develop and show understanding. Although interpretation of visualizations and other information is clearly critical to learning, becoming proficient in science also requires learners to develop many representational skills" (2011, p. 1096). Ainsworth summed up her beliefs about the importance of providing training in visual thinking and drawing to science teachers in the following statement:

> We need to research the fundamental mechanisms of drawing to learn. What skills do you first need to develop in order to best take advantage of learning by drawing? Perhaps some topics are sufficiently difficult to draw that attempting to do so is counterproductive... A further important research area concerns how teachers can best support their students to use drawing alongside writing and

talking in the classroom. However, what is clear is the growing interest in drawing as it reflects new understandings of science as a multimodal discursive practice, as well as mounting evidence for its value in supporting quality learning (p. 1097).

Recently, Tytler, Prain, Hubber, and Waldrip (2013) published the results of an extensive seven-year study on the effects of student-generated representation drawings on "engagement and conceptual learning, and capacity to coordinate representations as part of conceptual understanding" (p. 19). Citing work by Greeno and Hall (1997), Kozma (1997, 2003), diSessa (2004), Jewitt (2007), Ainsworth (2008), Tang and Moje (2010), and Hubber & Tytler (2010), Tytler *et al.* (2013) convincingly argued that throughout the process of constructing visual representations, semiotics (meaning-making mediated by symbolic representations) and epistemology (the practice of knowledge construction) were 'necessarily interdependent' (p. 69). The most salient findings from the Tytler *et al.* (2013) study (as pertained to my research) were that all participant teachers agreed that the representational approach made "new demands on their teaching skills and knowledge" (p. 25). In particular, teachers in their study expressed a need for the following: 1) more skill and practice in how to teach form and function relationships through drawing; 2) instruction in how to themselves create representational drawings; 3) instruction in how to move students from representational to symbolic, abstract drawing; and 4) instruction on structuring social negotiations of different student claims and justifications (p. 25). Teacher participants in *Project Draw for Science* embraced all four of these challenges as they engaged in capacity building training for teaching science through drawing.

An informative study by Finson and Pederson (2011) described and defined "visual data" and their utility in science education. They cited work by Esrock (1994), who argued that image schemas structure all our perceptions and cognitive acts and therefore, "have consequences for all forms of human understanding" (p. 112). They theorized that working with visual data was primarily the work of artists (p. 71) and acknowledged that a "comprehensive understanding of a problem and potential solutions to it are possible only when one knows not only particular skills in visualization, but also in knowing the right questions to ask" (p. 71). Agreed, but where is the *heart* in this study or the others? I invite you to discover what can happen when the cognitive and affective dance together, and hearts and minds function as one—a synergy made possible when science (or any other subject) is aesthetically explored.

An actor who does not know his or her "lines" or who has not taken the time to deeply study the personality and background of the character being portrayed is not likely to deliver an engaging or well-received performance. Many science educators (especially those having to teach out of field like Crystal) lack deep knowledge of their content matter. It's a scary sate of being and a problem in the practice of science education which I believe teaching through drawing can very effectively help address. Early on in my research, I came across a powerful science lesson planning model by Windschitl *et al.* (2013) called the "Big Idea Tool", a central component of which requires the teacher to write out a "causal explanation" of the science phenomenon he or she is going to teach, together with an explanatory diagram, all the while grounding the content in the prior experience of the learners. What I most appreciated about the Big Idea Tool was that it required a teacher to honestly explore his or her subject and pedagogical content knowledge. By using the Big Idea Tool, teachers can gather *evidence* as to whether they know their subjects well enough to make basic explanatory drawing models of the content they are required to teach. By correcting the deficits in their own subject matter knowledge, they can markedly decrease the occurrence of teacher-generated science misconceptions and improve the quality of their teaching performances—especially for those content areas in which they previously lacked deep knowledge and had therefore been reluctant to teach.

PCK AND THE BACK OF THE NAPKIN TEST

The construct of pedagogical content knowledge (PCK), or the means by which a teacher transforms subject content knowledge into meaningful learning experiences, has been the focus of much effort in teacher professional development (Solis, 2009), and is at the forefront of discussions of best practice for culturally and linguistically diverse students (Boutte & Hill, 2006; Ladson-Billings, Perry, Steele, & Hilliard, 2003). As a new teacher, I learned firsthand that it's almost impossible to meaningfully transform subject matter you don't know well into something that meaningfully connects with your students' heart and minds. The "P" in PCK is all about Performance.

All teachers in this study reported that developing the capacity to teach science through drawing afforded them with a very powerful means of improving their own content knowledge and made them better teachers, better communicators. These outcomes did not occur spontaneously, however. My team and I worked together for an entire school year. It is generally acknowledged that change takes time, which is why professional development instruction that does not allow time to practice adaptations so often fails (NPAET, 2003). Teacher belief systems and deeply engrained practices can be difficult to change. Guskey (1986) confirmed that if teachers can be convinced to use a procedure and actually see improvements in learning outcomes or engagement, they will demonstrate significant conceptual change. Driver (1989), Borko and Putnam (1995), and Scott, Asoko, & Driver (1992) have all proposed that *cognitive dissonance* must first be achieved if true conceptual change is to occur. To this end,

I created *The Back of the Napkin Test*. The name of this formative assessment was inspired by the important work done by Dan Roam (2008) in the area of visual thinking and problem solving. I now invite you to participate in an enactment of this diagnostic, for which you will need two dinner napkins (or paper towels) and colored markers:

> Ready? I would like you to write on the top of your first napkin the area of science you *most* enjoy teaching. For example, my napkin would have the word chemistry written across the top.
>
> Now put that napkin aside and write on the top of the other one the area of science you least enjoy, even *hate*, teaching. My own napkin would still have "physics" written at the top:
>
> Now, I'd like for you to pick up your first napkin again, and, using your markers, make a simple drawing explaining some area of your favorite content area. *Fieldnote: All teacher researchers in the study completed this activity with relative ease and apparent confidence. Most drawings involved plants in some iteration. All would have happily continued embellishing their drawings had I not re-directed them.*
>
> Okay, now will you create an explanatory drawing on your other napkin for your least favorite science area? *Fieldnote: Teacher participants immediately began scowling and even sighing. The most "hated" subject seemed to be physics. No surprise there. My own PCK in physics has a hole in it the size of the Milky Way. The room was deep with dissonance. Some gave up altogether. I called time quickly, having clearly made my point. I worried they might feel I had tricked them into exposing their ignorance.*

In the middle of the *Back of the Napkin* testing, I realized I might have made a big mistake to put my teachers through this exercise, especially so early in our work together. They were markedly disquieted by the experience. I was witnessing a tectonic level cognitive dissonance. I tentatively and gently tiptoed into the reflection phase of the inquiry by posing the following questions:

> True or False—Being able to make a simple explanatory drawing of a science concept can signify that you have a fairly deep understanding of that concept.
>
> How can creating explanatory drawings enable both you and your students to identify misconceptions about science content and practices which you may have?
>
> Why do you think you identified a particular *science* area as your least favorite?
>
> During what part of the school year do you teach your least favorite area, and how many days do you dedicate to its teaching and learning?

Their answers were highly revealing and consistent. "Because I don't have enough content knowledge in physics," one offered. "Because I think ecology is boring," announced another, crossing her arms. "I confess, I wait to teach simple machines 'til the last week of school, right before end-of-course tests," another sighed. "I HATE rocks!" said one who had filled her napkin with angry black blobby rocklike shapes. "Hmm," I replied. "This conversation is making you pretty uncomfortable, isn't it?" They rolled their eyes at me, waiting, somehow knowing it was about to get even worse. I took a deep breath. "Here's the big test of your commitment to this research," I announced: "I would like to challenge each you to develop lesson plans through drawing which serve the dual purpose of deepening your content knowledge in your weakest subject area as well as improve the chances of achieving meaning making with your struggling students." The wait time before I received *any* reply was alarmingly long. Definitely, I had pushed them too far too soon.

Would they freak and flee? After what seemed like forever, one by one, they agreed to give it a go. *Project Draw for Science* would become a reality after all. I gave them a break, and when they returned, I gave them all copies of their science standards, asked them to group themselves by grade level and to collaboratively begin to create inventories of drawings they should be able to quickly draw in order to explain these content matter areas. I asked them to also include canonical symbols from the accepted science lexicon such as standardized formulae for molecules and scientific/ mathematic equations whose meanings students would be expected to understand. They attacked the assignment with relish. They were smiling again, too. Whew!

EDUSEMIOSIS

Tang and Moje (2010) have argued that as students progress through learning sequences by generating a series of drawings, they will enact something akin to Peirce's "infinite semiosis", with one representation replacing another as new knowledge is constructed. (p. 82). Peirce's infinite semiosis theory was strongly congruent with that of his contemporary Charles Darwin's (1859) theory of natural selection and evolution. John Dewey's (1938) pragmatic philosophy of education, in turn, was clearly informed by both Peirce and Darwin's writings. Dewey (1938) characterized as *educative* and *democratic* those learning experiences which permitted students to acquire knowledge *naturally* through the process of ever-changing interactions with the environment. Specifically, Lemke (2004) has called for multimedia literacy in *science* education that is not limited to language or text. A multimodal pedagogy is not only more democratic but less likely to misrepresent science as *only* a tedious collection of facts to be memorized. Science taught *aesthetically* is characterized by a high regard for individual freedom, kindness, improvisation, and leading to growth by stimulating curiosity and individual initiative. Tochon (2013) has called on scholars to identify a "grammar of teacher actions" (p. 96) which signify best practice in edusemiosis. In this book, I do not presuppose to offer up any "rules" for teaching science aesthetically, but instead have represented multiple ways in which an aesthetic *syntax* may be employed to construct highly meaningful science learning experiences.

3 "KNOW"TATIONS AND THE NATURE OF SCIENCE

Creating Narratives

Few science teachers with whom I have worked have been afforded any training in the graphic art of visually organizing text and images on a single page. Often their notes are splattered across their classroom white boards like symbolic vomitus. Student versions of these notes are often even more confusing and certainly do not invite later study. Why? Because there is no story—nothing to make the scribbles (strange symbols written in a strange language) *mean* something.

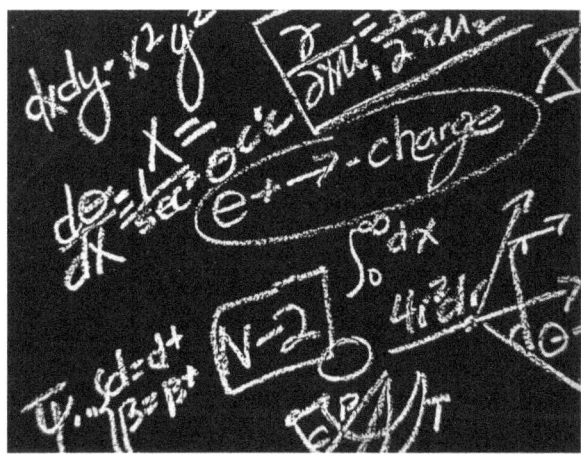

In Chapter 2, I cited Wisely's (1994) model of effective communication. Noted as key components were the "sending"

medium, the fields of experience shared by the sender and receiver, as well as the channel by which the knowledge was delivered. A "disconnect" in any of these features lessens the chance that the intended meaning is relayed. The science teacher is tasked with successfully communicating appropriate canonical symbol systems of the discipline so that students can acquire improved fluency in the language. To address the problem of breakdowns in communication between science teachers and students, I developed a versatile PCK tool which I came to call the "Know"tation, a single page drawing composition which visually captures the *gestalt* of a science teaching process or learning goal.

The form the "Know"tation takes depends on the learning goal which has been established and which stage of the 5E science inquiry learning cycle (Bybee et al., 1989) during which it is used. For example, creating an *Explore* "Know"tation can challenge the student to visually tell the story of an experimental investigation (thus making it a part of their own field of experience, not just the teacher's). Another *Explore* "Know"tation might include a series of generative drawings created as an engineering problem is solved. An *Explain* "Know"tation might be a student-created teaching/learning poster, diagram, serving as a set of "visual notes" created by a student to represent his or her synthesized understanding of a given body of informational text. For example, as part of test review on plate tectonics, I tasked groups of students to teach through drawing each of the different kinds of plate interactions. Each group was given specific content and vocabulary to graphically explain. Through performative drawing, we collaboratively and collectively worked through errors in understanding, and I could clearly discern those areas where there were gaps in their learning. In an earlier arts-based tectonics lesson, students choreographed plate to plate interactions through dance.

An *Explain* "Know"tation can be a very effective means of jointly synthesizing a big picture view of a phenomenon.

Regardless of how they are positioned in the inquiry learning cycle, "Know"tations graphically facilitate the formative assessment of both teacher and student learning. "Know"tations make visible misconceptions which can be considered in the moment and then revised, either individually or collaboratively. In this role, the *Back of the Napkin* drawings were examples of *Evaluation* "Know"tations.

In conceptualizing this play on words, I placed parentheses around the prefix "know" to remind one that the very nature of science involves a constant re-testing and revision of what we *think* we "know" based on the evidence we may have at a given time. As part of my "Know"tation instruction, I included the fundamentals of basic page design, including instruction on typography and layout, flow, use of white space, arrows and other techniques for visually organizing images and words on a page so that one should be able to "read" the story of what was happening on the page (or other drawing surface). I stressed that each science word chosen should be absolutely necessary for meaning making (otherwise, omit it). During a pilot study for this research, conducted with a group of high school biology teachers, I modeled the making of a "Know"tation, which I scribed while one teacher facilitated a lab using an Elodea plant "performing" photosynthesis:

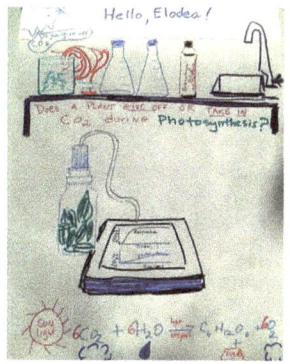

In fairly short order, these teachers were soon practicing creating their own visual explanatory stories ("Know"tations) as a central part of their lesson planning process and teaching performances. I stressed that even if they believed they had no drawing talent at all, their students would appreciate their willingness to reveal their own vulnerability. Following is an excellent drawing narrative performed by a biology teacher (who was very nervous about drawing in front of anyone) about the subject of diabetes, dead foot included:

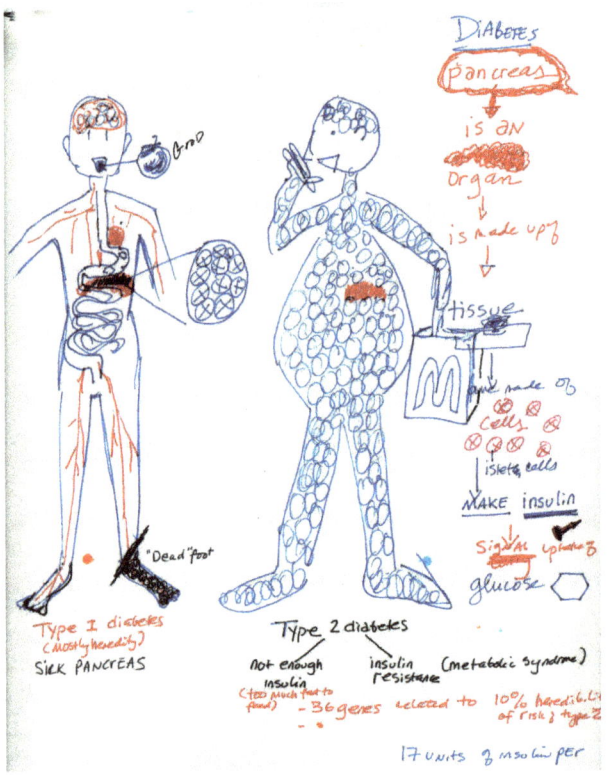

This teacher made especially effective use of common cultural metaphors, including the bag of MacDonald's fast food to represent junk food being eaten by the obese figure and the apple being eaten by the healthier looking form (thus linking students' fields of experience to the content). Her marker color choices were deliberate and directed the audience to "hot spots" of content knowledge. Two students of one of the teachers in this pilot study had the following to say about learning science through drawing:

While taking the test, I pictured the drawing in my head and that helped a lot.

I know that it helped me because everything that I drew that was on the test I got right but everything I didn't draw I got wrong.

An important learning goal for my capacity building course was to effect what Eisner (2002) called the "slowing down of perception" as a way of "noticing the world" (p. 10), something both teachers and students do far too little of on any given day, and especially *not* in the classroom. Eisner believed that the arts served the highly cognitive function of allowing us to *see* what has been said. He called *seeing* an *achievement*, rather than simply a task (p. 12). I found this sentiment to be strongly evocative of Pestalozzi's methods of teaching drawing. To effect a conscious act of seeing and maximal slowing down of perception, I taught all teachers in my study how to create blind contour drawings, whereby they drew the contours, or edges of an object or picture, without looking at the drawing in progress Most were very surprised by how well they could draw when they turned off their inner critics.

Next, in order to engage both sides of their brains and to yet again require them focus on being fully present with their work, I guided them through several experiences with ambidextrous drawings:

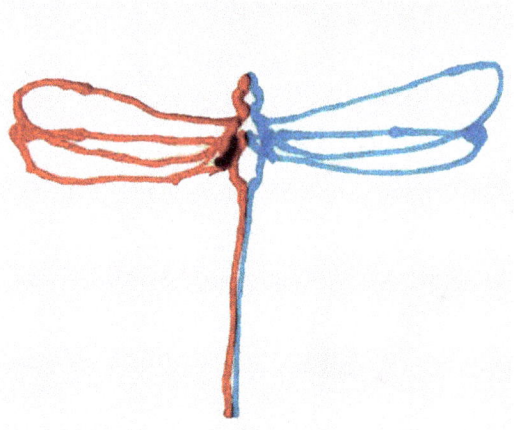

Teachers reported that students found this particular drawing exercise to be highly relaxing and to improve focus. One teacher started inviting students to create such drawings before summative tests. Finally, we worked on drawing the human form by studying a simple wooden manikin:

The making of explanatory drawings requires both teachers and students to synthesize key elements of science knowledge and/or tell the story of a science learning experience. Consider this student's "Know"tation about an investigation on energy transfer, especially the creative way in which the student represented the temperature data:

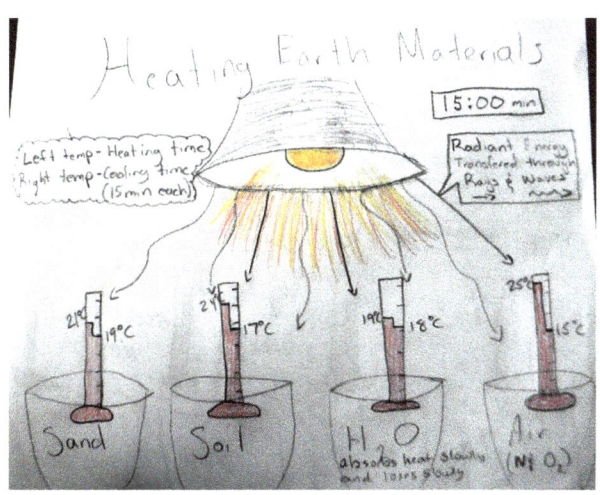

...Or this by this struggling reader about reptilian egg laying behavior. Iguanas were his "favorite animal":

MINDFULNESS

What is happening inside the brain as semiotic synapses are being laid down? Psychiatrist, researcher, and 2000 Nobel Prize laureate, Eric Kandel, discovered that "when we learn, we alter which genes in our neurons are expressed, or turned on" (Doige, 2007, p. 220). In the process, the genes make new proteins that alter the structure and function of the neuron. Kandel's team demonstrated that when we form long-term memories, neurons not only "change their anatomical shape [but] increase the number of synaptic connections they have with other neurons" (218). By proving that the brain is "plastic", Kandel, in my opinion, raised the responsibility bar for all educators. When we teach, we are actually shaping our students' brains by influencing which genes in our neurons are transcribed, and thus, what proteins are synthesized. Kandel's contemporary, psychiatrist Susan Vaughn, has likened teaching to "talking to neurons" (Doige, 2007, p.221).

Further reading in neuroscience led me to the American Psychological Association's online journal, *Observer*, where I found an article on the 1992 discovery of "mirror neurons" (Ehrenfeld, 2011), which I will summarize as follows: In 1992, a team of researchers at the University of Parma, Italy, working with macaque monkeys discovered that the same brain cells fired when the monkeys performed a task (like picking up a peanut) as did when the monkeys watched a person do the same task. They named these brain cells "mirror neurons". The article goes on to say that a few years later, neuroscientists found mirror neurons in humans and have subsequently learned from MRI imagining that mirror neurons fire even when a subject simply thinks about a particular task. Of significant interest, these mirror neurons showed up in the lobe of the brain most responsible for memory, the median temporal lobe. Scientists believe that mirror neurons are also responsible for much human empathy. If, for example, you observe that someone is sad or upset, the same neurons for these emotions will be triggered in your own brain. It makes sense that these mirror neurons evolved in humans so we could learn from observation and communication. It also makes sense to me that students are surely learning things teachers may not be intending for them to learn—like the misconception that science is not visual, nor creative, is inherently difficult to understand, and of little relevance to their day to day lives.

Can the mirror neuron phenomenon explain why *aesthetic* experiences are so *meaningful*? I think so. However, the research is so new that there is as yet no unanimous consensus. Here is what I posit: The human brain doesn't *light up* for just any experience. At best, it may "flicker" a bit during an *exceptional*, albeit unidirectional lecture; however, when the arts are infused into the learning process, neuronal *flares* should go off. In short, multisensory experiences should activate multiple regions of the brain. Lectures do not an aesthetic, transformative *learning* experience make. *Artistic* teaching and learning *performances*, however, are quite another organism altogether. An appreciation of what the arts can do for education should become normative teacher education practice. Virtuoso teaching performances are not required. As Godin (2012) has reminded us, something only becomes art when it is shared, that is, when the art-maker dares to make him or herself vulnerable. When teaching art is honest, albeit technically "flawed", it evokes empathy and a shared sense of humanity, making the experience interactive and meaningful. Should we not dance just because our hamstrings are tight or we occasionally trip? It does not matter if your science drawings are not "masterpieces". Even the simplest drawing can signify great meaning as learning is performed through the point of a pencil or even a child's crayon.

HAROLD AND THE PURPLE CRAYON

Again and again, teachers with whom I have worked in professional development settings have expressed gratitude for being guided through a performative narrative drawing technique I have been developing over the last two decades. My own personal fascination with the power of narrative drawing began with my childhood discovery of Crockett Johnson's book *Harold and the Purple Crayon* (1955), which conveyed the idea that your imagination can take you pretty much anywhere if you give it permission to do so. Armed with a single fat purple crayon (the kind with virtually all American kids first experience the magic of drawing), Harold drew himself a world full of fantasy and excitement. As a child, I was enthralled when I watched television artist-storytellers draw their stories in real time. Soon after completing my master's research at the University of Hawaii on teaching science through the creative arts, I began to experiment with enacting science content through drawing, creative drama and storytelling, employing a "live-action" drawing "performance" of science phenomena, all the while inviting my students to draw with me as the visual story "developed" (sort of like a photograph in a darkroom). I was intentionally applying the tenets of Alan Pavio's pictorial superiority effect and dual coding theory. Briefly, Pavio postulated that the brain stores information in both verbal and visual forms. After extensive research, he concluded, "the pictorial image code may be qualitatively superior to its verbal code" (Pavio & Csapo, 1973). This so-called 'pictorial superiority effect' has been extensively studied by Shepard (1967), Nelson, Reed, and Walling (1975), and Stenberg, Radeborg, and Hedman (1995).

I came to perceive the guided narrative drawing *not* as a kind of visual "lecture", but as an artistic medium through which understanding and learning could be negotiated. As my students and I drew together, we visually explored the narrative of science phenomena. Often, before tests, I would ask groups to collaboratively create drawing explanations of specific content areas and then bring them up to perform their drawings at the board. Not infrequently, we would also act out science processes, adding the power of embodied imagination to the learning going on. As audiences, we were to be respectful and encouraging of those putting themselves out there. In a spirit of congenial and collaborative learning, we would work through any misconceptions that still remained and offer constructive affirmative criticism. In every way, we were in these teaching/learning performances together, transformed in our being and by our deepened awareness and understanding of science matters.

An Intentional Undertaking Designed to...

4 Nurture

Performing an Ethic of Care

Peter London (1989) has called on all those who seek to bring out the artist in another (as I was trying to do with my teacher participants) to "be there fully" for those in our charge when they make the significant effort to share themselves in some way with us…"If it is transformation we want, this full living requires full meetings, full support, and utter concern" (p. 85-86). My proposing to science teachers that they become *artists* themselves is often a hard sell. The idea can seem positively threatening to a teacher who believes he or she lacks any artistic talent whatsoever. London (1989) described those stages through which those unfamiliar with creative encounters were/are likely to pass before any kind of transformation occurred: Typically, a novice will experience a "period of disequilibrium", which brings with it "varying degrees of stress" and a "defensive attempt" to preserve what is possessed and known" until gradually they "soften their hold" on old systems of beliefs about themselves and can begin to experience a "series of encounters and self-discoveries" through which they will "create their own personal meaning of art" (p. 79). Like a proud parent, I was filled with joy to see these teachers and students soar.

WHEN MIA SPEAKS

I don't know who was more surprised by what was happening in Marsh's class—he or I. In just over five pages of transcription of our very first Skype interview (conducted after he had been experimenting with teaching science through drawing for six weeks), I coded for thirteen separate *epiphanies*, each revelation reported with exuberance and delight: *It's been really amazing! They're not using their Ipads. Everything they're drawing is related to content. Wait 'til you see the details in the leaves...veination patterns, the stems, the labeling...beautiful! They're getting photosynthesis like you wouldn't believe!* I could hardly wait to visit Marsh's classroom. I had not even known he was training a new student teacher in the techniques as well, and she turned out to be a phenomenal addition to the study, as you will see. Marsh's most profound epiphanic moment occurred when he was able to use drawing to communicate with a previously withdrawn, practically mute, completely unsuccessful student named Mia (name changed):

Mia is a sweet girl.
She rarely says
>anything.

Mia was failing
all of her classes.
I met with her parents.
They were very
>distressed.

They just wanted her to be
>successful

I asked to meet her
after school.
I sat down across from her
at a table,

took out a sheet of paper,
and just started drawing a
funny fish.
She didn't say
anything, but she did
smile.

Suddenly, she picked up her
own pencils and started
       drawing
these amazing figures—
Manga style children with
huge, all-seeing
eyes.

After I caught
my breath,
I said,
Mia, let's make
       a portfolio.

I am going to ask you
some science questions.
Will you draw the
answers for me?

And Mia drew
what Mia knew.
She drew an illustration of
the levels of classification—
exactly what she had
       failed on the written test.
And then, she really
surprised me. I know why

the leaves change colors,
she quietly announced.
Yes! (she continued)
smiling.
The **red** is always there.
You just
     don't see it.

During classroom observations, I personally observed Mia's true colors emerging— smiling and laughing in class, actively talking with her classmates, and she has more than passed subsequent tests. What became very apparent during this research was that teachers who create caring connections with their students improve the likelihood of positive transformations and growth occurring in each of them. In this case, Marsh cared enough to engage Mia in a drawing dialogue, which, in turn, allowed her to feel comfortable enough to speak, for the first time ever, about what she had learned both in the classroom and through her own independent study. Mia clearly knew more than she had ever told any *teacher*, having dutifully played the role of the "dumb student", fulfilling the low expectations that had been established for her year after year. By valuing her drawing ability and conversing with her through appealing visual language, Marsh managed to break through Mia's deeply entrenched self defense mechanisms so that she might experience both academic achievement and self-actualization in the form of aesthetic capital. Thus liberated, she applied herself to traditional paper and pencil tests as well, and for the first time, began to earn passing grades in science.

Saldaña (1999) has directed the qualitative researcher to call forth what the character says about his life or objectives as well as what he has done to achieve those objectives. In their initial ontology statements, teacher researchers reflected upon those factors which had shaped them as they became teachers themselves. Marsh had previously shared with me that he was dyslexic and that the sense of alienation and inferiority he had experienced in many classrooms because of this disability were what motivated him work extra hard to ensure his dyslexic students acquired agency in his classroom. I asked him how teachers had made him feel inferior. A tumble of words came out, which echoed Lortie's (1975) seminal study of schoolteachers:

>These teachers make
>no effort to
>DISCOVER
>what kind of
>learner
>you are.
>They try to
>explain in
>the only way
>they understand or
>have been
>taught.

BONES DIGS FOR DATA

Two months into our research, Bones became Marsh's student teacher. He insisted she also develop ways of teaching science through drawing. Bones, a numbers-oriented, very logical individual (whom we named after the television forensic scientist) agreed, but under the condition that Marsh allow her to determine herself whether teaching science through drawing would have any impact on the D.U.E.'s. (District Unit Evaluations). The D.U.E.s were a set of standardized multiple choice science content questions sent over from the district office and administered to all students on their Ipads. Within ten minutes after student submission, teachers received the graded results. Marsh dreaded the D.U.E.'s and worried that his poor readers were consistently failing them. He also shared the almost overwhelming oppression he felt trying to "stick to the pacing chart"—a prescribed science curriculum also externally created in the district office. He said he "just couldn't do it. It depressed him." It depressed me as well to witness such a gifted teacher so stifled. And then Bones entered the research study and greatly cheered us both. In addition to fully embracing the teaching of science through drawing, she also asked to experiment with using creative drama in her teaching of science, all with the purpose of generating quantitative data. What a gift she was!

I was struck by the confidence and poise of this young undergrad as well as her graceful ease in communicating with middle-schoolers. During one of my observations, she skillfully corralled them into quickly settling down to take their D.U.E. on arthropods, their second Science Through Drawing lesson. Marsh and Bones had several ELL students in their classes, as well as many reading below grade level. During one of my observations, Bones was reading the test to one boy, who was functionally illiterate. Regardless of his reading disability, he had clearly *learned* a LOT about arthropods. I listened as he talked about his drawings while considering what answers to choose: *I don't remember drawing that word*, and *well, we drew that real big, so it must have been important. I choose answer "B". That's right! said Bones. Way to be!*

After the quiz, Marsh and Bones called the students together in the large open space in their classroom, which had been cleared since my last visit. I learned that Bones had suggested moving the lab tables to the periphery of the room so that they could have what she called "Town Hall Meetings". Her innovation created a living space for a democratic educative experience, strongly reminiscent of those described by Dewey (1938) in *Experience and Education*. I marveled at what happened next, as twenty-five students eagerly clustered on the floor in the center of the room:

Marsh: I'm really proud of you. We have today's quiz results. Let's talk about the data we're gathering about learning science through drawing, because you know that Bones and I have been doing an experiment to see how this works.

Bones: Okay, now I want everyone to activate your thumb gauges. Remember, thumbs up if you're happy about something, sideways if you're so-so, and thumbs down if you're sad or unhappy about what I tell you. I want you to think back to the sponge and worm quiz. How do you feel like we did on that?

Twenty-five Students: Hold up a "mixed review", based on thumb gauge directions. Most gauges were in the neutral, so-so position. Two were up, and seven were down.

Bones: Okay, now let's move on the echinoderms quiz. How do you think we did on that quiz?

Students: About thirty per cent of the thumbs were up, about half at neutral. Only three were down.

Bones: Now, for the quiz we just took on arthropods.

Students: Nearly all thumbs were pointed up. Only one was pointed down.

Marsh: Let's look at this data together. On your sponge and worm quiz, the average was a 71. We weren't drawing then. On the echinoderms quiz, after we started drawing, the average was 83. Today, the average on the arthropod test was again an 83. We're not there yet, but this is still a big increase from when we weren't using drawing. Do you think it helps us? We want to know what you think.

Bones: Why do you think you did so much better on the arthropod quiz, which you did!

Student 1: Well, you know during that first three weeks of school, all you were doing was *talking* ...You kept *talking* about inquiry, but I had no idea what you meant.

Student 2: But now we're drawing *and* acting!

Student 3: And taking visual notes.

Student 4: I like getting to choose my own colors.

I could feel my own pulse quickening in response to the aesthetic energy sparking off these kids. Then Marsh announced they were transitioning into yet another way of exploring science through art:

Marsh: Now, I am going to invite you stand on the shoulders of my most favorite artist, Michelangelo. (He held up a book about Michelangelo's work). Michelangelo was most famous for his sculptures, but before he could create them, he made many detailed drawings. Bones and I want you to create a sculpture of the invertebrate of your choice, beginning with making a preliminary drawing.

Marsh then directed them to the supply table, where he had modeling clay all ready for them to use. Around the room were baskets of specimens and a wide assortment of nature study books from which to find drawing references. Many students were eagerly looking up invertebrate pictures on their Ipads. Within only a few minutes, incredibly representative sculptures of invertebrates began to materialize around the room.

I tore myself away from interacting with the young sculptors to interview Bones, who was finally freed up for a few minutes. I asked her about her first impressions with using drawing to teach science:

Bones: Well, I always understood the importance of integrating science with other subjects like math and social studies, but I had never thought about integrating it with art until I started student teaching with Marsh. He got me to watch some of your teaching videos and your lesson on "Phylum Picture Pages" and right away, I knew I wanted to try getting my students to draw their notes.

Merrie: What was the first Science Through Drawing lesson you developed on your own?

Bones: The arthropod lesson. Marsh let me teach the whole thing. What I found out was that drawing motivated me to spend more time with the content than I would have before. It's changed the way I thought science should be taught and how I plan to teach.

Merrie: Can I see your lesson planning sketches?

Bones: Sure, here are the first two pictures I created for the kids, on arthropods.

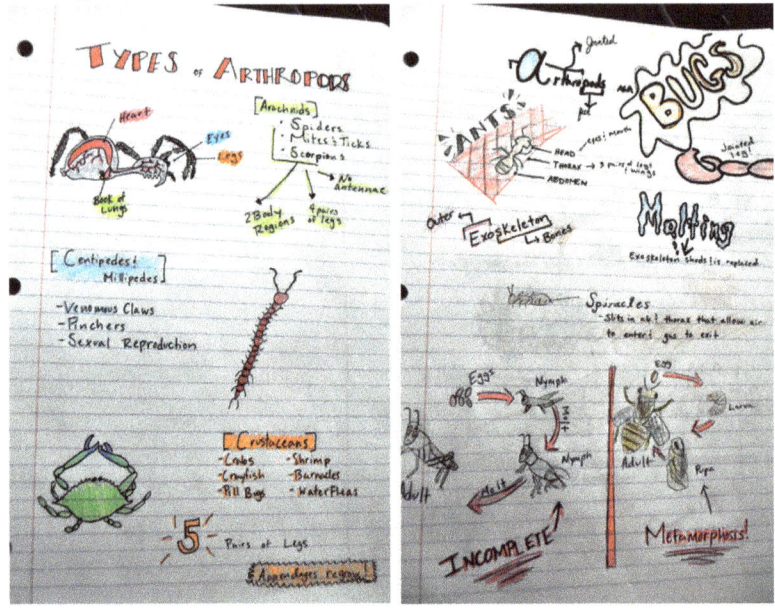

And here are my mollusk pages.

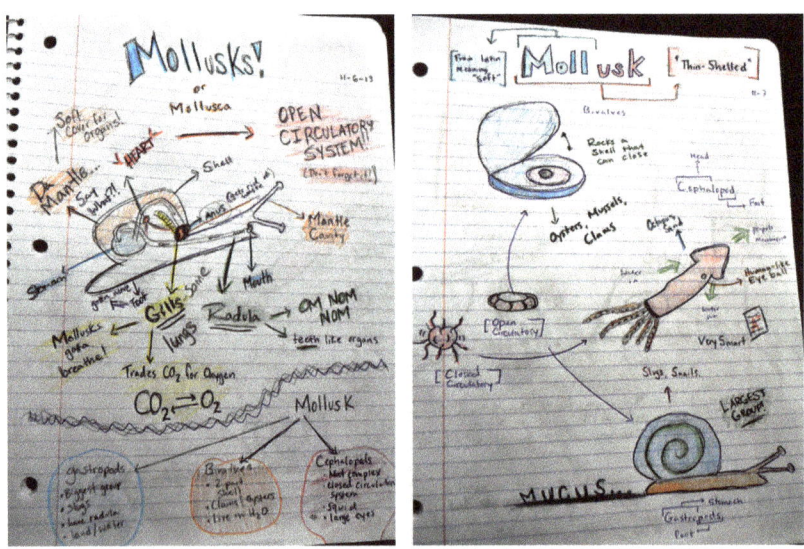

Merrie: Wow, you have a great eye for page design!

Bones: Well, my dad is a graphic artist, so I grew up around art. Just like you did in your video, I taught the kids about creating their own fun typography, and using different fonts for main headings vs. subheadings. I also told them that I was just modeling *one* way of designing the page, and that I wanted them to add their own unique artistic flair—to make it their own. Some kids did end up pretty much copying my design, but others went out of their way to put their signature creativity on the work. I also got them to check each other's work to make sure the content was accurate. The science still *does* have to be there!

Merrie: What students do you think are benefitting the most from this approach to teaching science?

Bones: Oh, for sure the ones who were failing before. But the gifted kids are really loving this, too, because it lets their creativity shine, and they're not so bored. But not every single student likes the drawing approach. We have one student who said she hated drawing, but loved acting. So while I'm drawing out stuff, I also have students act out life processes like metamorphosis. No one is forced to draw. We also use videos for things like crabs molting, which have really helped the kids see what happens in nature. We're trying to find as many ways of connecting science to what the kids care about as we can.

Merrie: You are already a master teacher, and you're not even certified yet! Can you give me an example of some of the creative drama you used?

Bones: Hey, Sadie, will you and Robert come over here a sec. Will you come act out metamorphosis for us? *(Bones explained that Sadie was an ELL student who struggled with science words and her quizzes before being invited to use drawing to explain what she had learned. Robert was dyslexic but very artistic. She reported that he was now passing his tests.)*

Sadie: Sure! Well, first we get on the ground and pretend we're eggs.

Robert: Then we turn into nymphs.

Sadie: Then bigger nymphs.

Robert: Then finally adults. This is incomplete metamorphosis, like a grasshopper.

They then got on the ground and start chanting Egg, Nymph, Nymph, Adult over and over, while changing the shapes of their bodies. Then Sadie announced that the bumblebee was different, and they started all over, this time chanting Egg, Larva, Pupa, Adult, while changing their movements and shapes.

Merrie: Wonderful! Bravo!

Robert: A couple of days ago, Bones let us pretend to be sea urchins. Did you know they poop out their guts when a predator tries to catch them?

Merrie: You didn't act that out, too, did you?

Both together: Yes, we did, too! *(Bones rolled her eyes and laughed. We all laughed.)*

Merrie: Robert and Sadie, why do you think learning science through drawing is helping you, because I know your grades have improved.

Robert: Well, um...I think it works, because, you know when you draw, you have a visual in your mind and it sticks better than when you just write it in words.

Sadie: With writing notes, I don't like it much...so I just forgot about them...I didn't pay any attention to them.

Merrie: Why do you think you would forget notes that you wrote down compared to notes you drew?

Sadie: Um, well, because they don't have a visual to them…when I have a visual…I can remember what's going on. *(She opened her notebook to a colorfully drawn rendition of Bones' arthropod notes).* See my arthropod picture page? :

You see that I wrote the word "jointed" to show what "arthro" means. Same for "exo" meaning "outer". See the foot here, how it's jointed? I have colored arrows pointing to this ant's head, thorax, and abdomen. Then I've written the word "exoskeleton" very large with an arrow to explain that it acts kind of like bones do to support the ant. I've written "molting" so I can remember what that looks like, too. Down here, in the corner is the picture of incomplete metamorphosis that we just acted out for you.

Merrie: I really like this work, Sadie! It looks like you've been able to put into a single page what might have been lots of written notes.

Sadie: Yes! That's what makes it a "Know"tation!

Merrie: How do these "Know"tations help you study for a test?

Sadie: Well, after class, I go back and color code everything. All the colors...they stick out in my mind...and I remember them during the test. I write my words in pink, purple, and blue, because those are my favorite colors, and that helps me to remember.

Merrie: That's amazing, Sadie! It's like you've created a mind map for yourself!

Sadie: Yes! *(Then the bell rang. I thanked Robert and Sadie for their time and again praised their work.)*

In her final paper for her student teaching seminar, Bones told the story of how she embraced a method of teaching science which she noted was not a "normal" one. What she most appreciated was the ongoing opportunity to easily and quickly formatively assess her students' understanding of the material she was teaching. What *I* most appreciated about Bones was her insistence on regularly checking in with her students by using the "thumb gauge" meters to assess their feelings, too about what and how they were learning. By so doing, she was greatly contributing to what Pestalozzi (1801/1898) would have characterized as "wholeness" of the child. She showed herself to be a master of joint collaborative pedagogy, and the learning products she and her students created together were exceptional. In her final paper, she also wrote about challenging her students to write cinquain poems about what they had learned and drawn and then uploading both the poem and the drawings to the classroom Edmodo site so that she and Marsh could formatively assess their understanding. From this information, she and Marsh were able to "determine which topic they needed to cover deeper and which topics the students were really grasping". In her final reflection, she wrote: *While the shifts in the grades were important, I was more impressed with how excited my students were about the information I was teaching and how it was being taught. They put so much effort into their work. I learned that if you stick with this method, even though it takes a lot of work, students will almost teach themselves. Students who develop curiosity about a topic begin to fuel their own learning and thinking.*

Indeed, the "shifts in the test scores" were impressive. Here was Bones' graph of her students' quiz grades *before* employing drawing (Worms/Sponges quiz) and *after* (Echinoderms/Mollusk and Arthropods quizzes):

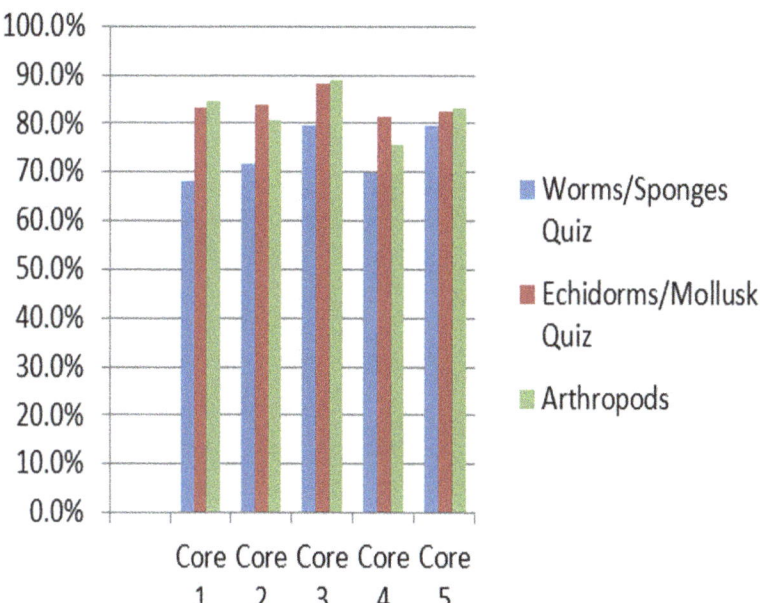

What lay behind these test scores can never be measured. The questions are, what matters most and for whom? The narrative and the pictures tell a far more complete story of the transformations that can occur when science teachers and their students draw to learn.

MS. MAYA AND MUHAMMAD

"Maya" is a sixth grade science teacher only recently certified. She used to be a construction contractor, but came from a family of dedicated teachers. She had refrained from entering the teaching profession because she had witnessed how draining this career of care became for both her mother and grandmother. But she claimed that the genetic tide was too strong, and she at last succumbed to its pull. She was born to teach and intentionally allowed herself to be "led into troubling spaces occupied by others" (Denzin, 2009, p.266). I hope that my interpretation of her performance does more than merely evoke empathy. I want Maya's story to inspire action and change. Recalling Conquergood (1985), I have sought (for the analysis of her performance data) to write text which "interrogates, criticizes, and empowers" (in Denzin, 2003, p. 55). I invite you to imagine a school deep in the rural South, where eleven-year-old Muhammad was in personal crisis until Maya became part of his lived experience.

According to Maya, *something happened to Muhammad in the third grade. He plummeted from the 93rd percentile in reading to the 8th. He started acting out. In the fifth grade, he received over twenty referrals to the principal's office for misbehavior.* His parents fretted about their beautiful boy. At the end of the fifth grade, he was diagnosed with severe number and letter dyslexia. That same year, a new sixth grade teacher arrived at their school, a freshly minted P.A.C.E. certificate holder in science. Her students called her "Ms. Maya". For the first time ever, some previously failing students began to demonstrate significant academic improvement under Ms. Maya's care. The principal was impressed and asked Ms. Maya if she would consider teaching all the sixth grade "low babies" the next year. They were mostly all boys, and Muhammad was among the lowest of the low in terms of his test scores. According to Maya, *my lower kids thrive on challenge. I think that's because no one has ever challenged them before.* Maya enrolled in my study, in spite of the fact that she had significant levels of drawing anxiety:

The shear idea of drawing in front of my students
made me cringe.
How was I supposed to teach science *and* draw
for understanding?

The standards had to be broken down into
meaningful information,
into pictures that brought it
to life.

I knew my principal would approve of
employing new strategies
to layer conceptual information.
My teacher Merrie's encouragement allowed me to
reach outside my fixed mindset and
approach the task which I feared.

I immediately signed up
not knowing where
she would take me as a developing
educator or what
the road ahead for my
students would look like.

When school started in August,
the last thing on my mind was ART!

My students had been labeled rude, disrespectful,
low and unmotivated.
That first day I met Muhammad,
I knew I had to change the way
science was being taught, not just for him,
but for myself.

Muhammad was not alone
in his struggles with comprehension;
there were other who I discovered were
struggling
readers,
writers
and decoders.
Initially, my lesson plans
were done entirely on
a pre-formatted template,
but
I realized after the first Skype conference
with my teacher, that
I needed to "draw" my lesson plans.

In order to teach through drawing, I had to teach myself
to embrace my fears and
buy into a
new way of thinking.

I began by adding color to my anchor charts,
something I did not use at all last year.
I incorporated basic arrows for directional flow,
not only for my students,
but for myself.

In keeping my journal
about the struggles of the class and
teaching my students to move away from

traditional notes and toward
drawing their notes, labs, and assignments,
I realized I was
deepening my own knowledge.

My students fought me in the beginning.
They did not want to draw,
they had never been allowed to draw
and unfortunately,
they had been told not to draw,
because that meant they were playing in class.

Before I could teach them about drawing,
I had to change their mindset about learning.
For so long they had been taught using
the drill and kill method,
not knowing how to apply their knowledge.

    After our first session on the basics of graphic page design, Maya immediately began organizing science content in "anchor charts" for her students on large *Post-It Note* flip charts, using directional arrows and key science vocabulary words, color coded for their "importance": Red/Orange words were HOT", signifying the most "important" science standard in the lesson. The Yellow words were *almost* as important, and the Green words were "extra stuff" they might see on the statewide science achievement test. Her first "Know"tations used words only:

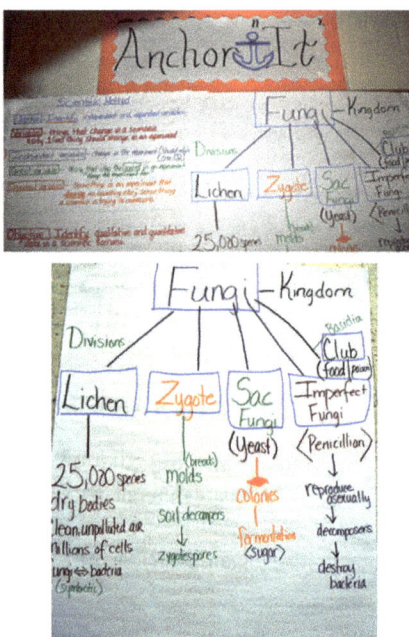

Very soon, however, Maya's "Know"tations began to feature a central communicative image and only the most necessary words. I personally was struck by how she had mastered the art of visual communication of science content:

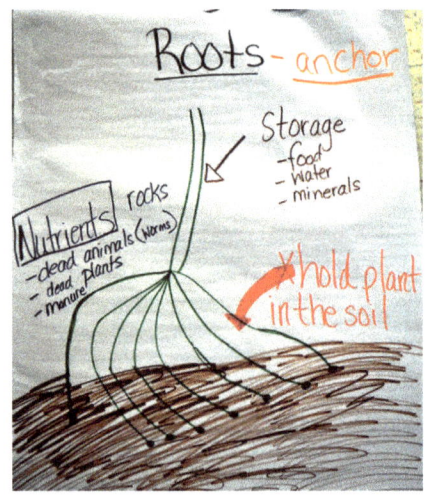

Maya's huge visual explanatory models were hung all around her room for easy access and review before summative assessments. An index card with the featured science standard(s) and indicators was attached to each sheet of chart paper. Maya reported that, over time, she became more and more comfortable drawing in front of her students, even though they made fun of some of her drawings: *Ms. Maya, that's the funniest mushroom we ever saw!*

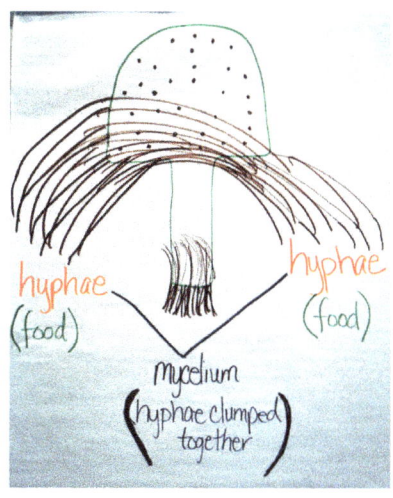

During a Skype interview, I asked Maya, a gifted writer, if she had tried using the visualization techniques we had explored in my training class. She immediately brought me into a visualization experience she had created for her students. They had been studying plant succession. First, she showed them videos of the Yellowstone National Park forest fires of 1988. Then she said:

> I want you to close your eyes.
>
> Now, imagine that awful heat is
>
> getting closer and closer to you.
>
> The flames are
>
> reaching their fiery fingers toward you...

At which point, several students shouted, "Oh, it's getting hot!" That's when Maya said she knew she "had them". She continued the story.

> Finally, finally, come August,
>
> The rains began to fall, and
>
> Mother Nature began her work
>
> Anew.

Maya reported that she dramatically stopped the story and asked her students to draw in their journals a picture of the new life that would soon appear in the forest as well as label any science words about plant succession that they remembered. To Maya's delight, words like "pioneer plants" appeared over drawings of mosses. There were also birds and squirrels and exploding pinecones, popped, said one student, like popcorn in the oven. Maya reported that the principal just happened to come in during this lesson and that she laughed out loud because she was so happy to see the "low babies" getting it. Maya was extremely protective of these kids, finding nothing "low" about them at all, except their self-esteem. She celebrated their diversity:

In order to receive maximum participation from my students,

> I hold them to the highest standard possible.
>
> I hold myself to the same high standards.
>
> The initial impression I want my students to receive
>
> is a warm, welcoming environment where
>
> they can learn how to prepare for
>
> real life.

Students with varying backgrounds can teach each other

about their differences

in parental involvement, social upbringing, and the

experiences they have had to that point in life.

I am a teacher who seeks to educate a

well-rounded student.

In order for this to happen,

I must myself be willing to

take risks.

This means continuing my own education to seek

fresh ideas on how to engage my students.

This is why I am willing to draw for them,

even though I am no good at it.

Maya was clearly enacting what Denzin (2003) has described as an "ethic of care and empowerment" (p. 105), making deep connections with her students, the ones everyone had given up on, as together they experienced the liberating freedom of learning by taking risks. Maya has fired the flames of success and succession, for none more so than Muhammad:

Muhammad: But, Ms. Maya, I don't want to draw any pictures. I'm no good at it.

Ms. Maya: I'm no good at it either, Muhammad. If I can do it, so can you. And remember, I'm not grading you on the quality of your art work—only on what you can show me that you know—and that's a lot! Here's the problem we have. Your dyslexia is keeping you from telling me with *words* what you know right now, even though I'm sure we'll figure out how to work around that. My teacher told me that we all remember things better through pictures, and that if we draw a picture of something and then write a word label, we'll remember the word, too. Here, look at this book my teacher gave me. *[Maya showed Muhammad a copy of Mona Brookes' Drawing With Children: A Creative Method for Adult Beginners. I have used this book for years, for it gently improves one's "noticing" skills, guiding the new drawer through progressively more complicated copying exercises, allowing students to improve their skills in visual discernment in a way that feels safe and non-threatening. When students discover how easily they have created very complex designs, invariably their drawing anxiety is quickly reduced and smiles replace frowns.]*

Muhammad: Here's what I drew last night, Ms. Maya. It's about how birds help seeds get planted.

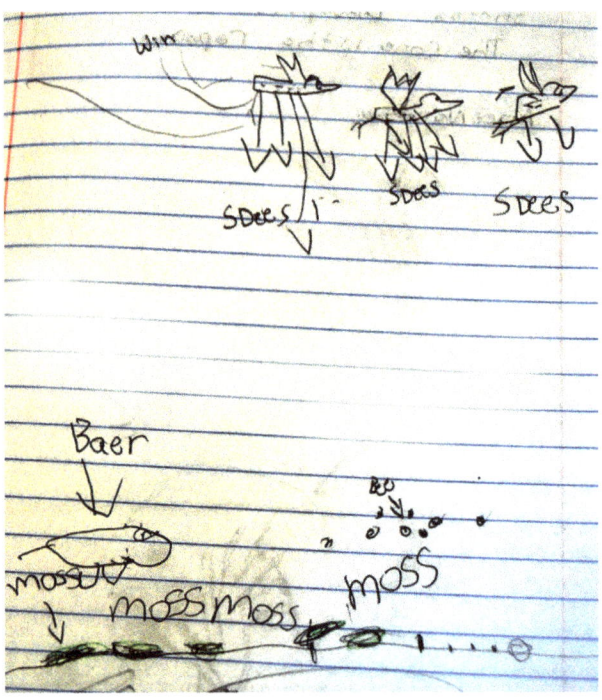

And here's what I wrote in my notebook:

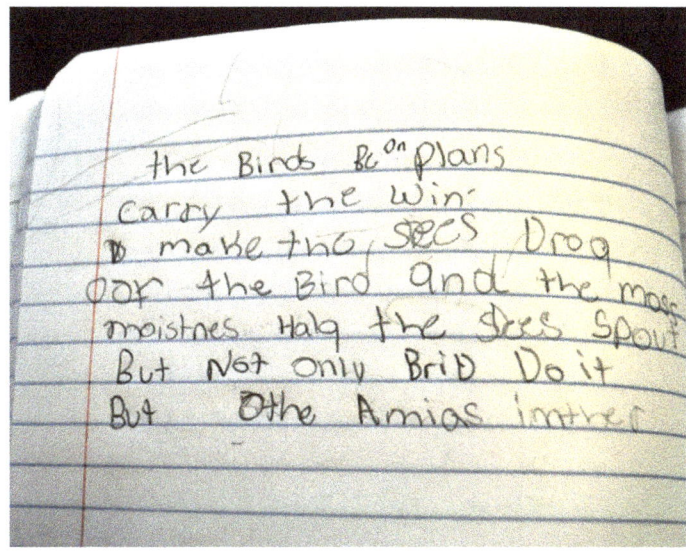

Ms. Maya: I'm really proud of you, Muhammad. You're really getting this stuff! You're the only one who learned that sometimes, even tadpoles hibernate if the water's too cold. Your "Know" tation says it all:

And I love this drawing here. Even though technically, a group of lizards are not called a flock, your drawing tells me that you know the meaning of the word:

Muhammad: Yeah, I love reptiles and amphibians.   Hey, look at this page I found for drawing frogs!

Ms. Maya: You know, Muhammad. Your computer skills are amazing. But, um, Muhammad, I hate to keep talking about it, but we still need to get your test scores up. You're not getting your vocabulary. I think the drawing is helping you, but not enough. Since you're so good on the computer, I think we should try creating some "Know"tations using an Ipad app called *Educreation*. My teacher told me about it.

Muhammad: Cool! You keep talking about your teacher. Who is she?

Ms. Maya: She says you can call her Ms. Merrie.

Muhammad: Did you tell her about me?

Ms. Maya: Yes, and she wants to meet you when she comes to visit next week!

Muhammad: *(frowning)* Oh, does she know I'm dyslexic?

Ms. Maya: Yes, she does.

Muhammad: And what did she say?

Ms. Maya: She said that she thinks it's great you're willing to try drawing pictures of what you know in science, even if you don't think you're very good at drawing.

Muhammad: People think dyslexia is a disease, like they can catch it or something. They think I'm stupid 'cause you have to read to me— 'cause when I try to ready for myself, everything looks jumbled. I have to work even harder than anybody—just to get the same grades as they do (*his voice starts to crack*). I don't want to be like this!!

Maya: I know, Muhammad. I can't imagine how hard it is for you. Did you know that some of the most gifted people in the world are dyslexic? You can be anything you want to be,

Muhammad

Muhammad: I want to be an engineer!

Maya: Then an engineer you shall be. All the more reason for you to learn to use new computer apps like *Educreation*. I'm new at it, too, so we can learn together. You can work on these after school with me and see if your science grades improve. We can start by creating a computer "Know"tation on animals. *[They both load the Educreation app onto their Ipads.]*

Muhammad: Okay. So...now what?

Ms. Maya: Well, let's type the main idea at the top of the page, just like we've been doing on the anchor charts. Then you can go online and find pictures about the different groups of animals we've studied.

Muhammad: You mean like carnivores, herbivores, and omnivores?

Ms. Maya: Exactly! Don't worry about the final design just yet. Go ahead and find a representative picture of each. *[Muhammad is very adept at web browsing, and immediately finds lots of animal images.]*

Muhammad: We need vertebrates and invertebrates, right?

Ms. Maya: Of course! I also want you to type in a brief description of each type of animal, too. Just do your best with spelling. Remember, what we want to try to produce is a single page that could be used to teach someone else about animals.

Muhammad: Like on the anchor charts?

Ms. Maya: Right! Remember the elements of design we talked about in class? You want to use as few words as possible, typed in readable fonts. You can use different colors of type to organize the information, too. The pictures should be arranged in a pleasing way with white space between them. Your eyes should move easily through the composition if you've been successful...

Over the next few months, Maya worked with Muhammad to create what they came to call "review slides", using the *Educreation* app. Here are two they jointly produced on Animals and Plants.

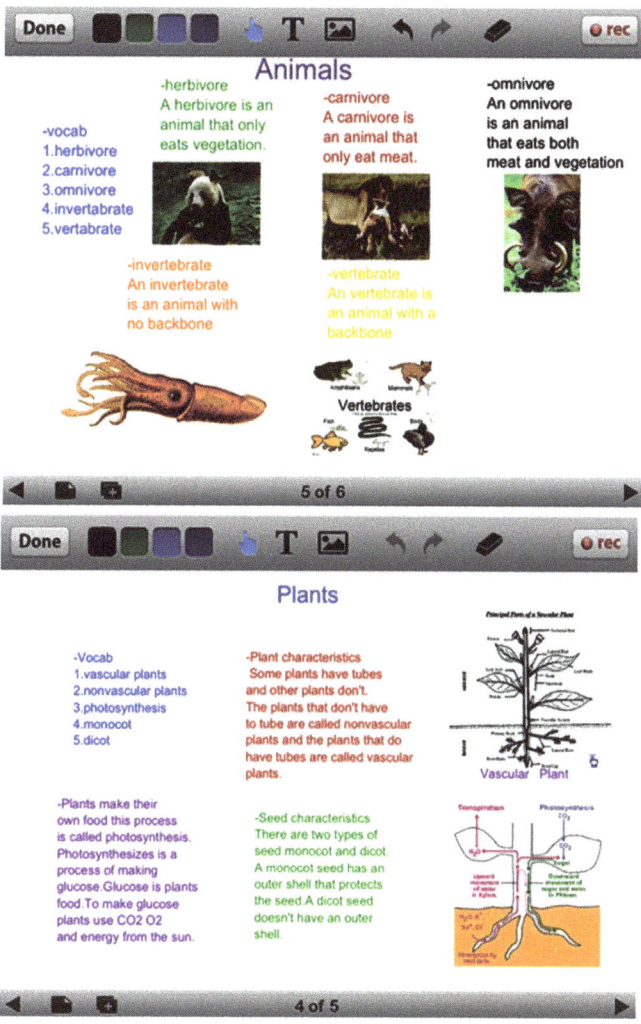

By Christmas, with the use of the *Educreation app* and the continued support and care from his parents and Maya, Muhammad's grades in science improved from D's and F's to A's. Encouraged, Maya suggested that the ELA teacher allow Muhammad to use *Educreation* to learn his vocabulary terms in that class, too, again combining pictures with the new words and definitions in Muhammad's own words. He continued to draw with Ms. Maya and the other students during class time. After only a few weeks, the ELA teacher burst into Maya's room, grinning with the news of his significant test score improvement. The ELA teacher reported Muhammad's progress to the principal, who was also a special education teacher. The principal then invited Maya to present her innovative use of the drawing app to a team of special education administrators at the district office. Maya was receiving significant support from her administrators as she continued her classroom action research on teaching science through drawing. The parents of her previously struggling students praised what they clearly recognized as "pedagogy of hope" (Freire, 2004).

By the end of the school year, Maya reported to me nearly across the board academic improvement on science standardized test scores for all eight of the students she had pre-identified before the study as struggling in science. Muhammad's science achievement test scores improved by three grade levels in nine months, from the third grade level at the beginning of the year to a "met" sixth grade level in the spring. When Maya asked him how and why he thought his test scores had improved so markedly, he replied, 'Because you cared about me. You drew with me, even though you weren't any good at drawing either.' You helped me do those *Educreation* slides to review for the tests." The learning outcome: teaching science through drawing in a caring learning environment had afforded both Maya and Muhammad (and the other previously struggling students as well) a wider semiotic field wherein meaning-making in science might be achieved and the capacity for achievement expanded.

DO MY SHELLS MAKE ME LOOK FAT?

When we last saw Crystal, she longed for a better world for herself and her students. In one of our earliest interviews, I asked her to describe how a perfect world would look:

> I want to live where
>
> I don't have to
>
> lock my car,
>
> lock my house,
>
> engage an alarm every night,
>
> and have a concealed weapons license.
>
> In a perfect world, the kids wouldn't pay
>
> for the mistakes of the adults in their lives.
>
> In a perfect world, I wouldn't be
>
> stuffing my kids' backpacks full of food on Fridays
>
> because there is no food at home.
>
> Their parents wouldn't be in prison and
>
> their kids would not die on the streets.
>
> In a perfect world, kids would
>
> not be crushed by illiteracy and violence.
>
> They would smile and laugh and
>
> love each other.

When I asked what role she hoped to play to make her students' lives better, she answered:

> I will push,
>
> and shove,
>
> and yell as loudly as I can
>
> so that they are not ignored,
>
> cannot be ignored.
>
> I will bring some delight to them each day, and
>
> tell them they are amazing people,
>
> no matter where they come from or
>
> where they will go.

Crystal was as luminous as a diamond, outwardly daring anyone to fracture her resolve to improve the lives of her students. Inside, however, her full heart was quivering. She was terrified of teaching science, but now found herself being required to teach all four core academic subjects in a self-contained classroom of special needs adolescents, most of them having autism spectrum disorder. What Crystal may have lacked in formal training in science education, however, she more than made up for with her super power of caring. Sweeping her fears aside, she purposefully strode into training with me because she believed that drawing might open pathways to understanding for her special needs students, where more traditional approaches had failed. At the epicenter of her epistemology lay the belief that the educational system had failed children like those in her care, and that they were most definitely not "expendable". Right out of the blocks (only three weeks into her classroom action research), she made sketchbooks for her students for their unit on lab safety. She had them dedicate their sketchbooks with the same inscription I had asked my teacher researchers to write in their own sketchbooks.

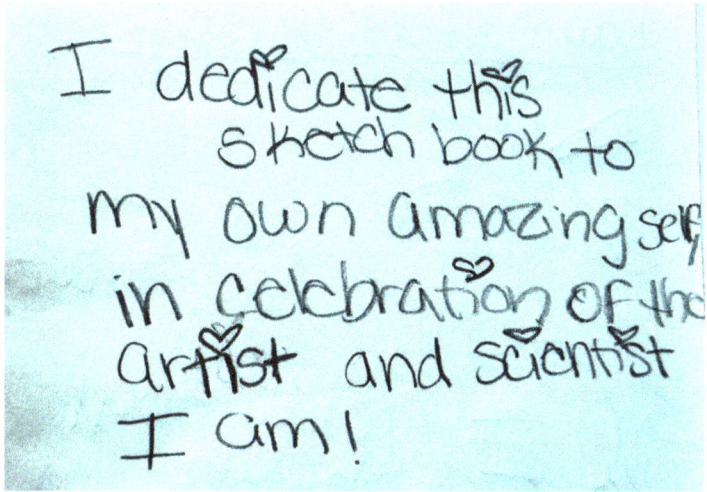

And then she invited them to create their own cover designs.

During one of our interviews, Crystal told me there was *not a typical kid in my classroom. For most, autism impedes their*

learning. Most lack appropriate social skills. Also, a lot of their parents have learning or psychological disabilities, too. Boys in special ed get into trouble. I keep them out of trouble and advocate for them. I can tell you that autistic kids hate being different! They don't want to stand out as anomalies. So, when we do the drawings, it puts them more on a level playing field. They're good at it, too! She described one autistic boy in her class whom we named Shelmore. According to Crystal, Shelmore *loves to draw, but needs to use an ink pen because the sensation of the pencil dragging across the paper makes him crazy. He's exceptionally bright and talks constantly on a wide variety of subjects. However, he lacks social skills. He irritates the other students because he won't be quiet.* She described another boy we named Timothy who has speech problems and who has to cover his ears when the school bell rings. A girl, Natalya, *reads at the sixth grade level but cannot comprehend a word she's read. She draws beautifully, however, and her handwriting is perfect.*

During an after-school classroom observation (the only time I was permitted to visit this school), I asked Crystal how she helped these special needs students connect with science content. Her answer again revealed a deep understanding of the power of aesthetics to engage the learner: *Well, when we talked about the open flame, we created a chant that went like this: Open flame—OUCH! And then we did an air drawing of the flame, just as you taught us to do in your class. They loved that! Then, I would say, "Draw your glove!" And they did. I was amazed at how much pride they all took in their drawings.*

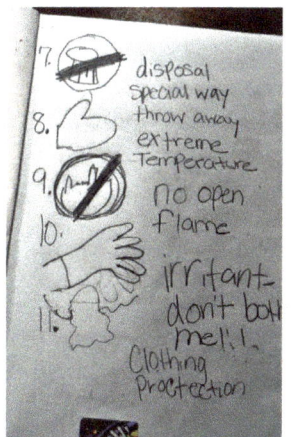
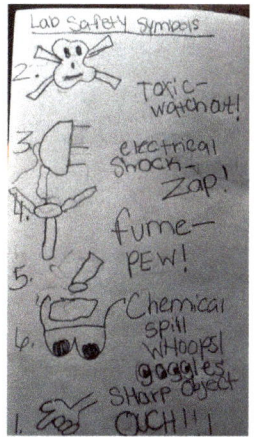

Crystal then told me about she helped her students develop relationship skills by evaluating each other's drawings: *I asked them to tell their drawing buddies one thing they liked about their partner's illustration and one thing they would improve. They were very considerate of each other's feelings. I've actually got them thinking in an evaluation mode; so they might make an observation about, say a graduated cylinder, that the measuring marks are drawn very well because they were so evenly spaced apart.*

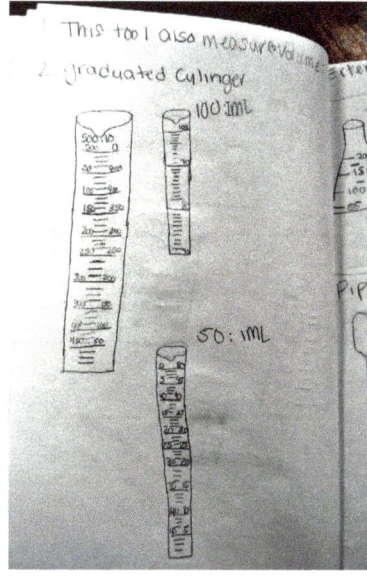

Crystal continued her metasemiotic reflection: *I believe the drawing builds collaboration and the idea that they are artists. This builds their self-esteem. What I am seeing is a comraderie developing around their drawing. They can share their pictures with each other, and they don't put each other down. Instead, they lift each other up. Autistic children have big social interaction issues, so for them to work together is a big deal. With drawing, autistic kids have a chance to succeed. I know for a fact my kids have never been invited to draw their understanding before. They all have reading difficulties. Ask them to find something written in a notebook, and they can't do it. But they remember everything about their drawings and can find the material again and again. Their sketchbooks give nonverbal support to what they might not be able to say:*

As part of her unit on lab safety and measurement, Crystal brought out assorted lab equipment for them to handle and to draw, a privilege they had never experienced as special education students. I found her approach strongly evocative of Pestalozzi's method of teaching orphans, who were not traditionally afforded an education in the 1800's. Anderson (1931/1970) has written the definitive treatise on Pestalozzi's quest to develop a "road to the fullness of life" for the "oppressed poor" a curriculum for the "natural development of the mental and physical powers of individuals, hitherto neglected or misdirected" (p. 3). Included in Anderson's analysis is a translation of Pestalozzi's most important work, *How Gertrude Teaches Her Children* (1801/1898), in which Pestalozzi explained his philosophy of education:

> The child must be brought to a high degree of knowledge both of things seen and of words before it is reasonable to teach him to spell or read…I am obliged to put aside that first plague of youth, the miserable letters; he must have nothing but pictures and things…Learn therefore to classify observations and complete the simple before proceeding to the complex…Every line, every measure, every word is a result of understanding that is produced by ripened sense impressions and must be regarded as a means towards the progressive clearing up of our ideas (Pestalozzi, 1801/1898, in Anderson, 1931/1970, pp. 3-55).

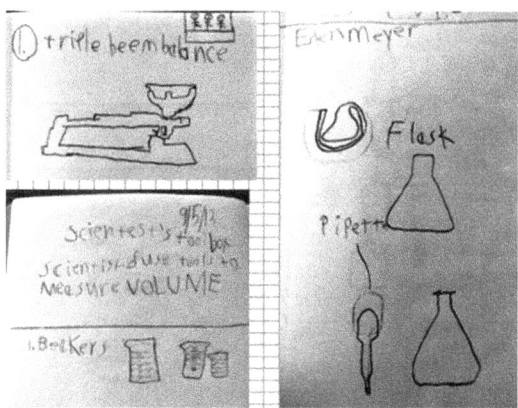

Crystal's aesthetic, sense-based approach to teaching science was strongly Deweyan, her approach reminiscent of his descriptions of the *educative* experiences he described in *Experience and Education* (1938). She fully recognized the importance of creating opportunities for sustained perceptual awareness in order to effect deep learning: *All the other kids get rushed through the curriculum and stay only at the surface level of understanding.* Crystal had very definite opinions about what is *not* best education practice: *You know, the way students are being taught contradicts the way the brain works and learns. Sixth, seventh, and eighth graders' brains are changing so fast we shouldn't be overwhelming them with so much new material. We should be helping them to acquire deep learning and tap into their imaginations. Drawing makes a student slow down so they can really see.*

I was at times nearly overwhelmed with emotion by what Crystal was doing with this research. The lessons our team learned from her about teaching autistic adolescents were invaluable. Somehow, she had managed to bring a group of students whom the system had closeted in a back room behind a closed door to a grade level or better understanding of science concepts. That she had managed such a teaching performance in *chemistry*, her professed "least favorite" subject in science, bordered on the heroic. Her extraordinary ethic of care was evidenced by her Herculean efforts to deepen her science pedagogical content knowledge in chemistry. Like the other teachers in my study, she had given up a week of her summer to improve her inquiry teaching skills. Crystal had signed up for the chemistry group. Even with the extra training, she confessed after our *Back of the Napkin* Challenge that she still felt overwhelmed by the challenge of teaching this subject. I just asked her to do her best. What that turned out to be was a show stopper.

Not only did Crystal learn the content, she developed a performative narrative drawing lesson in which chemistry concepts became characters in an engaging story. Planning, designing, and executing a jointly produced narrative teaching performance is an advanced application of teaching science through drawing. It requires extensive content knowledge and much time to develop. In all my years, I had never before witnessed a beginning science teacher even attempt it. By raising her teaching craft to the level of art, Crystal transformed her own hard work and risk-taking into significant aesthetic capital for her students, resulting in transformations which, according to Uhrumacher (2010) can cause a student to undergo an entirely new way of being in the world. I was wowed and not a little humbled. I was clearly in the presence of a master. Here was an out of field science teacher imaginatively communicating some of the most difficult content in either middle or high school science, and to autistic students, no less! These students had been deemed by most at her school as being incapable of learning at such high conceptual levels. In the middle of our interview, Shelmore entered the classroom. He marched right up to the desk at which Crystal and I were sitting. Crystal introduced us, and I told him I had heard a lot about his drawings in science class. He was rocking back and forth from front to back foot, grinning widely. Crystal asked him if he'd like to tell me the story of "Andy the Atom" by showing me the Smartboard images they had saved after they had drawn them in class. He was more than happy to oblige, and eagerly performed the beginning of the story for me:

During the entire performance, Shelmore was beaming, clearly enjoying being given the opportunity to show a visitor how much he knew about chemistry. I found myself grinning and was enthralled. He clicked on the next few slides and continued his performance:

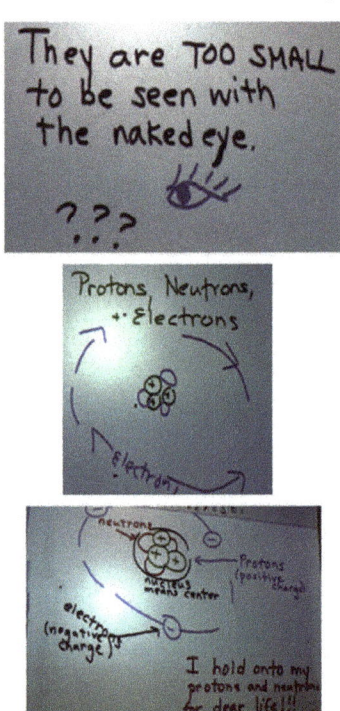

Here was a student whom the school system, according to Crystal, believed could not achieve at grade level, especially in a subject as "hard" as science. Instead, he was confidently explaining extremely abstract concepts like electron shells. Even more amazing to me was that Crystal had dropped chemistry in college and had learned all this subject matter content on her own. Her determination was astonishing. Shelmore continued the narrative, clicking on new slides as he progressed:

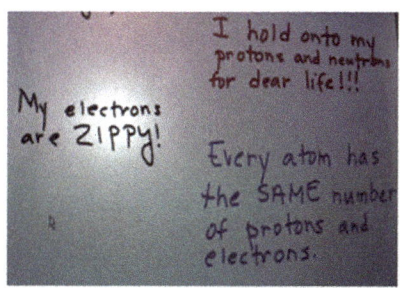

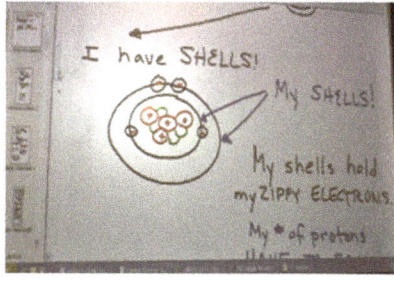

Suddenly Shelmore giggled. "I like the next slide the best":

We all laughed out loud. What fun we were having! I then asked Shelmore if he'd show me his drawings of the story of Andy the Atom. He said, "Okay, but my mom is waiting." Crystal asked him to choose two pages to show me, which he did, again reading the words with great enthusiasm, especially the part about his electrons being zippy!

Academicians have come to understand that drama, make-believe, and role-play in the science classroom (not just in the elementary school) can be a vital part of conceptualizing the nature of science (Aubusson, Fogwill, Barr, & Perkovic, 1997; Author, 2009; Barab & Duffy 2000; Dorion, 2011; McGregor *et al.*, 2012;). Dorion's (2011) extensive literature review has contributed greatly to the disrupting of negative stances against the use of anthropomorphic analogies to teach science. He cited the work of Taber and Watts (1996), who determined that even undergraduate level science students defaulted to anthropomorphic explanations, in which "cause is assigned to desires and perceptions (e.g., atoms *want* to form bonds) as part of their scientific reasoning" (p. 1). Crystal's strong background in psychology and frequent successful use of role playing in her teaching enabled her to intuit the kind of connections with self and content that a performed, anthropomorphic visual enactment of extremely abstract science concepts might afford:

I *know* that I am teaching better

because

I have to try to first

picture the concept

myself

and then

break it down into

pictures

and

conversation.

This means that

I understand it better—

am less likely to

'jump steps'—

I can more easily *see* and then

remedy

misconceptions.

My autistic students—

who often cannot

express themselves in a

timely, coherent manner—

can with *pictures,* tell me the main idea of concepts in a sequential and more fluent manner than during just regular conversation. Through drawing, they can be themselves.

5 *Notice What There Is to Be Noticed*

Inside and Out

Master art educator, Kimon Nicolaides (1891-1938) declared that drawing has little to do with talent or technique, but rather with the act of coming into "physical contact with all sorts of objects through the senses". He urged his students to imagine that, as they were drawing, their eyes were touching what they saw. In this way, drawing could become an extension of two senses at once. He reminded his students that mistakes were an integral and natural part of learning. In the process of making revisions, the drawer would "find out facts for themselves" instead of being "limited the rest of their lives" to facts the instructor related (xiii). Nicolaides's drawing methods had everything to do with actively *seeing* and almost nothing to do with passively *looking*. Very few people spend much time *seeing* the natural world anymore—not even children. A few years back, I had the pleasure of hearing Richard Louv (2005), author or *Last Child in the Woods: Saving our Children from Nature Deficit Disorder* speak at an ocean education conference. What I immediately realized was that my own research was about a related kind of NDD—Noticing Deficit Disorder. I soon thereafter created a series of cartoon drawings depicting how I thought James Thurber (another of my favorite authors) might have viewed NDD of both kinds:

The third drawing reflects my hypothesis that there would likely be associated with NDD (of either kind) higher incidences of both carpel tunnel syndrome and neck pain associated with too much *indoor* and sedentary computer use. I would now like to introduce you to Sky, whose students are most definitely *not* suffering from NDD.

SKYBEAMS

Upon entering Sky's class when it's in session, one has the sense of having stepped into a magic circus arena. A skeleton assembled from recycled plastic containers and assorted science themed mobiles bounce on stretchy bands from the ceiling. Student-created posters and projects fill every centimeter of wall and perimeter table space. Two hedgehogs squeak from a corner. A central lab table is piled to the overflowing with multicolored papers, scissors, glue, pipe cleaners, and piles of science magazines. There is a bright pink bike helmet with springy purple antennae protruding from it. There is no place for the eye to rest without encountering yet another promise of possibilities. For the purposes of this research, I named this teacher "Sky" because she was so "out there" and uninhibited.

> My name is Sky.
> I am easily bored.
> If I'm bored,
> my students are
> bored.
>
> I love experimenting with
> new ways of
> teaching.
> I'm also not afraid of
> making a mistake.
> My students know I
> am human, too.
> My principal trusts me
> to teach the way
> I think is
> best
> for my students.

In one of our early sharathons, Sky told Marsh that she personally couldn't survive if *she* were forced to adhere to a curriculum pacing chart as he was being required to do. "I have to try things out," she declared. "I teach on the fly. Thank goodness my principal doesn't mind that I'm ADD!" We all laughed, but then agreed that some of history's greatest achievers are believed to have been "ADD". Depending on what source you read, Einstein, Leonardo da Vinci, and Thomas Edison would have likely all been put on Ritalin today. How tragic it would have been to have subdued these over-the-top creative people with sedating drugs. Sky's astute principal recognized that her boundless energy, enthusiasm, and creativity (her so-called ADD) were what were made her such a valuable asset to both the school and the students. He applauded her voluntary participation in *Project Draw for Science*, asked for frequent updates, and asked her to present her action research findings at both school and district meetings. Her students' parents also loved this new way of teaching science and said as much. Throughout history, the arts have flourished when supported by patrons. Sky responded by eliciting an astonishing number of science through drawing artifacts—her own as well as her students. I have included here only a tiny fraction of her considerable output.

About a month into our research, Sky emailed me to say she was experimenting with using the narrative drawing technique to teach about plant and animal cells—but with a new twist. She had come up with the idea of allowing her students to make huge drawings of cells on their desks with washable markers. They had loved the approach. Just as we had practiced in our capacity building class, Sky said she had created a numbered drawing "script" of her own to follow. As soon as I could find someone to keep my dogs, I made the three hour trip to her school—just in time for the test review on cell organelles. The students were ready to begin the moment I arrived:

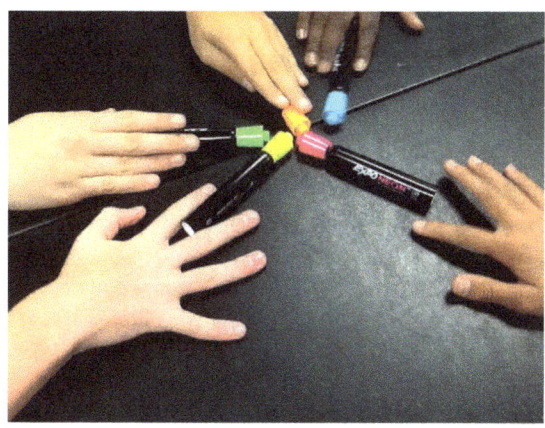

    Sky and her students took turns drawing, she on the white board, and they on their desk tops. A few times, she intentionally drew the wrong thing, saying something like "Did I draw this right?" Each time she was quickly corrected by alert students: "No! Animal cells don't have cell walls, and they're not so square looking either!" Sky negotiated the learning process with questions designed to assess her students' understanding of organelle form and function. She encouraged all tablemates to inspect each other's drawings and to offer help if a seat mate were struggling.

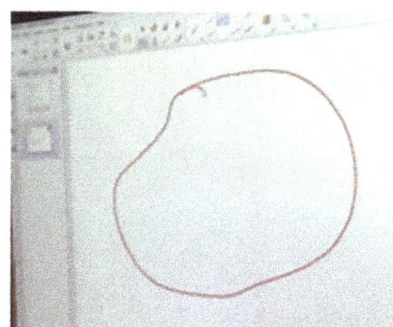

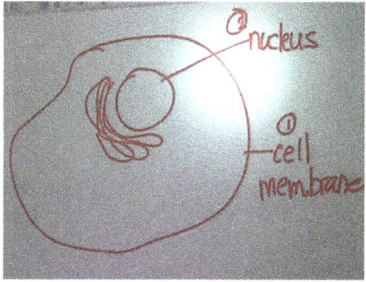

When Sky was confident that her students understood the form and function of each cell organelle, she created an opportunity for them to advance (according to Anderson's 2001 revised version of Bloom's taxonomy of educational objectives) from a factual to a more conceptual level of knowledge acquisition.

She challenged groups to build models of amusement parks in which the "rides" were analogous to the role of specific cell organelles. In order to build these science models, students were required to think at the cognitive level of analysis, evaluation, and creation. Many also incorporated a keen sense of humor:

Next it was time for Sky to incorporate procedural knowledge into her microbiology lesson. This time, her action research goal was to determine whether using drawing could help her students understand the relationship between magnification and field of view (FOV)— how much of the specimen can actually be seen in the visible image field. She reported that teaching this concept had always been difficult for her and that she never been able to adequately explain the phenomenon with words.

This time, without telling them why, Sky asked her students to make careful drawings of prepared onion root tip and cheek cells at two different magnifications. She wanted them to discover for themselves the inverse relationship between magnification and (FOV). She handed out jar lids so that her students could trace circular templates to represent the FOVs and instructed them to draw exactly what they saw. As part of her own lesson planning,

Sky eagerly created her own cell drawings and emailed them to me:

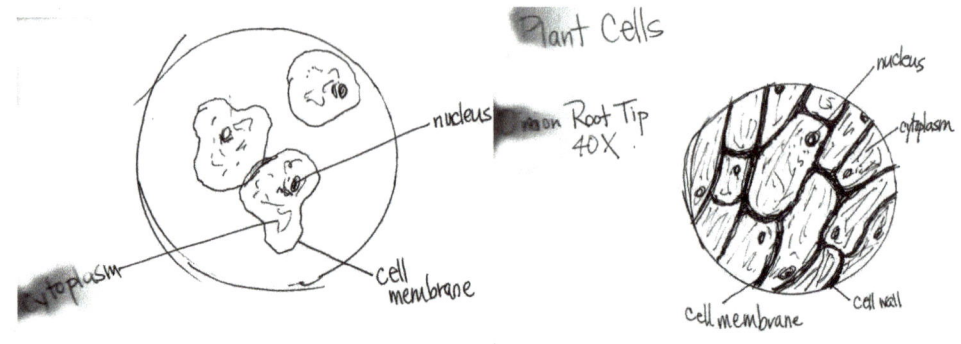

Within a few days, I received pictures of these student drawings:

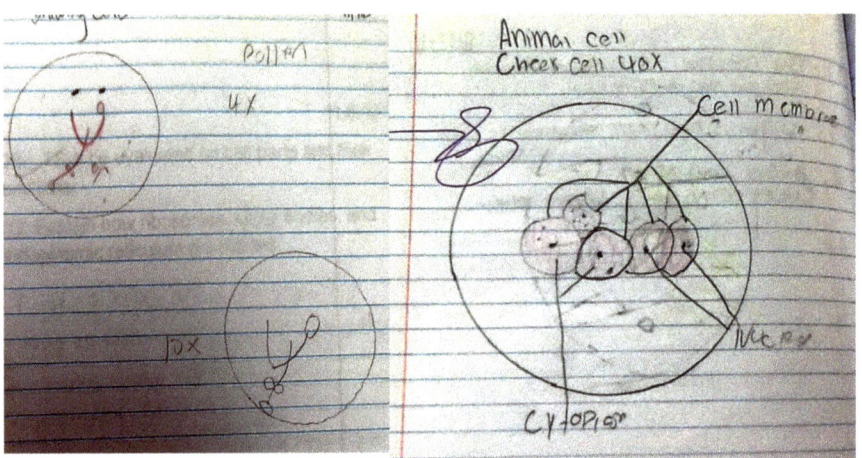

Sky reported that for the first time ever, her students "got" that the greater the magnification, the less of a specimen you could see under the microscope. Drawing to *know* had worked!

Now that Sky had experienced "success" teaching microbiology through drawing, it was time for her to take on what she had identified as her "least favorite" area of science—ecology.

During her *Back of the Napkin* test, Sky had created the following "explanatory drawing" about ecology as a subject. "NF" denotes "Not Favorite".

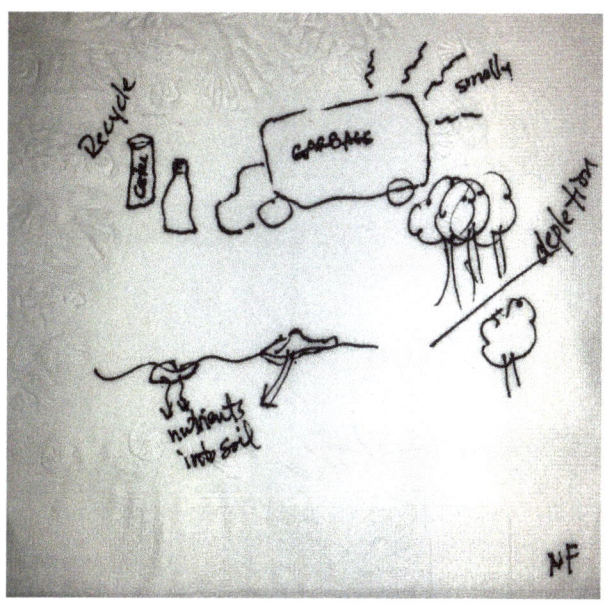

At this juncture in her inquiry and metasemiosis, Sky associated the subject of ecology only with smelly dump trucks, recycling, some vague reference to "depletion" (rainforests perhaps?), and soil nutrients. Now determined to deconstruct her entrenched dislike of the subject of ecology (a major part of her required teaching standards), she began anew, starting with trying to identify the "Big Ideas" (Windschitl *et al.*, 2013) associated with the ecology standards, which fell under the umbrella standard of "Matter and Energy". Here were her initial sketchbook planning entries:

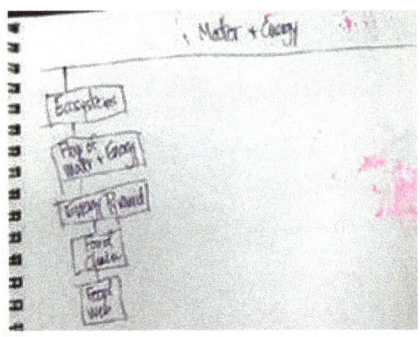

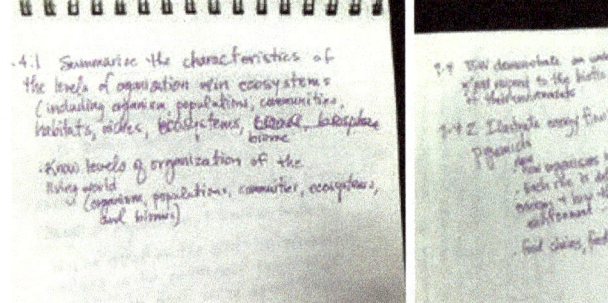

Among the learning indicators were *levels of organization within ecosystems, energy flow in food chains and food webs,* and *population dynamics such as limiting factors carrying capacity, niche, mutualism, commensalism, parasitism,* and *predation,* as well as the *interactions between biotic and abiotic factors.* Sky selected the latter as the overarching "Big Idea". She then had an epiphany, which was quickly followed by a pedagogical innovation: Sky's principal encouraged teachers to bring students outside. In fact, the school had two learning gardens on site. Unencumbered by the obstacle of having to teach about nature *inside* a classroom, Sky decided she would take her students out on the school's nature trail and ask them to *notice, draw,* and *describe* in their sketchbooks all the abiotic and biotic features they could discover, as well as ways in which these features interacted with one another. By way of a summative assessment, they would be required to synthesize their findings to create a small booklet of "Know"tations about Abiotic and Biotic Factors, using the following instructions:

"Know"tations of Biotic and Abiotic factors – Nature Trail

1. Create a booklet using 3 pieces of white legal paper (staple on the crease – top, bottom, middle). **Teacher must create a booklet AND do whatever the kids have to do!**
    a. Add Title, headings for each page, and page number.
    b. Front Cover – Title: Interactions between biotic and abiotic factors.
    c. Head the back of Title page (p.1) Dedication – this is where they dedicate their booklet
        i. "I dedicate this booklet to my own AMAZING self in celebration of the artist and scientist I am!" Sign their name
    d. p. 2 "Biotic Factors"
    e. p. 3 "Abiotic factors"
    f. p. 4 "Interactions between biotic factors"
    g. p. 5 "Interactions between biotic and abiotic factors"
    h. p. 6 "Interactions between abiotic factors"
2. Grab a clipboard and a pencil and head out to nature trail.
3. Outdoor classroom #1 (benches)
    a. Have students close their eyes and just LISTEN (2-3 minutes)
    b. Open up to p. 1 and start describing what they heard. Then students will share some of those things.
    c. p. 2 Students list as many abiotic (non-living) factors; water, soil, air, temp, sun

that are present outside on the nature trail.
   d. As we walk along the trail students are to NOTICE other biotic and abiotic factors. Teacher and students can share some of the different "stuff" they notice.
4. For time purposes we skip outdoor classroom #2 (observation deck)
5. Outdoor Classroom #3 (Meandering stream and soil horizon erosion area)
   a. This site provides students with some different biotic factors that they may not have noticed before. (Snail trails along the bottom of the creek, crayfish towers, soil horizons, etc.) Add the new observations on the correct page
   b. This is where students begin to illustrate what they have written down on p. 2 and 3
   c. Teacher provides crayons and colored pencils. Students are to find a quiet spot and focus on illustrating what they have written down on both of those pages.
   d. Allow however much time you can but not be late for the next class.
6. When back to class collect the booklets before dismissing students.

Once out in the woods, Sky observed that her students became both transformed and transfixed. One boy, who lived in the inner city, declared that this was "the happiest day of his life". He had never experienced such beautiful surroundings:

Sky first led her students to an eroded stream bank, where she knew there was a well-defined soil profile:

She introduced her students to features of stream banks and asked them to make representative drawings of the soil profile as well as any other abiotic and biotic interactions they observed:

They threw themselves into the task, utterly enchanted with being invited to notice the natural world. They discovered that there was distinct beauty in a stream bank and that embedded in the exposed soil horizon was the story of streams past, which they captured in their drawings:

Suddenly, one student squealed with surprise, having spied a tree branch crawling with little "bugs", busily scurrying over the surface of a golden, mushy mass of something.

Sky recognized that the "bugs" looked like aphids and that the tree was a beech. That was the extent of her knowledge of the phenomenon playing out before them. Never without her cell phone, however, she Googled "aphids and beech trees", and to everyone's delight, was able to pull up an image of "beech blight aphids", whose body shapes were remarkably similar to that of the quivering mass before them. Its scientific name was no less alien sounding: *Grylloprociphilus imbricator*.

    Sky reported that she touched the branch with a stick, and it came alive! Shrieks of delight and/or horror escaped from each student. After quickly reading the reference she had found online with her cell phone while standing in the woods, Sky explained that what they were seeing was one of the coolest *symbiotic* relationships she had ever witnessed and that the beech blight aphid feeds on the sap of the American beech. Reading from her phone, Sky explained that beech blight aphids exude kind of sticky honeydew in which the spores of the black fungus *Scorias spongiosa* take hold and then build up huge colonies. None of these goings on seems to harm the tree much, if at all. (http://mhelm.com/blog/beech-blight-aphid-grylloprociphilus-imbricator/). The sweet honeydew, in turn, attracts all kinds of insects like bees and yellow jackets, as this student noted in her detailed observations of these interactions.

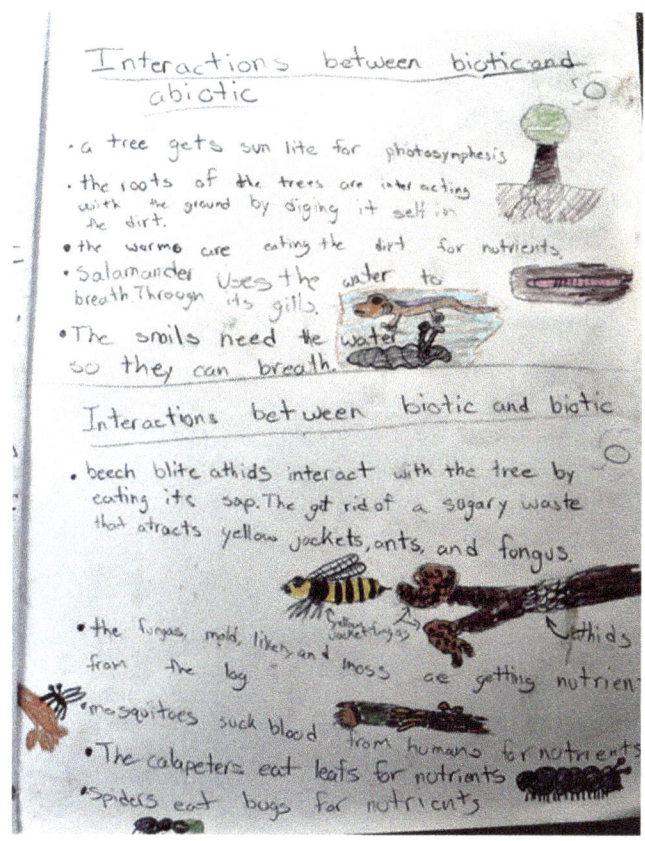

Sky reported that while standing there while her students drew and described the beech-fungus-aphid interactions, she had another epiphany—she would ask her students to create an ecology mural, which depicted the interactions. In the meantime and while she worked up the lesson plan and rubric, they would return to class to complete their Abiotic/Biotic booklets in class.

Much to their delight, Sky even invited her students to draw their learning about ecology on the floor of the classroom with the same dry erase markers they had used for their desk top cell drawings. Her innovative ways of teaching a subject she had only recently condemned as "boring" were nothing short of astounding. Her investment of aesthetic capital had paid out dividends, which she was not about to squander.

She immediately got back to work and developed sketches of how a "scene" from this imagined ecology mural might look. Had I seen her drawing before she presented to her students it as a model, I would have been able to help Sky avoid inadvertently teaching a misconception about nitrogen cycling. However, our team would have missed out on a very important learning lesson ourselves. In graphical representation, the arrow icon has impactful signifying power. Sky's drawing (below) gives the incorrect impression (via the arrow directions) that nitrogen (N_2) is *directly* taken in by leaves. There is also a misleading note stating that soil bacteria "break nitrogen down to be returned back to plants". More accurately, specific types of soil bacteria can convert nitrogen from the N_2 form into ammonia (NH_3), which all organisms can use to synthesize proteins and nucleic acids. (These are really difficult abstract concepts, which I don't think I grasped until college.) During a recent conference at which we were presenting data from *Project Draw for Science*, Sky shared how her own misconceptions ended up in her students' drawings. All I could think was what an act of courage this was on her part.

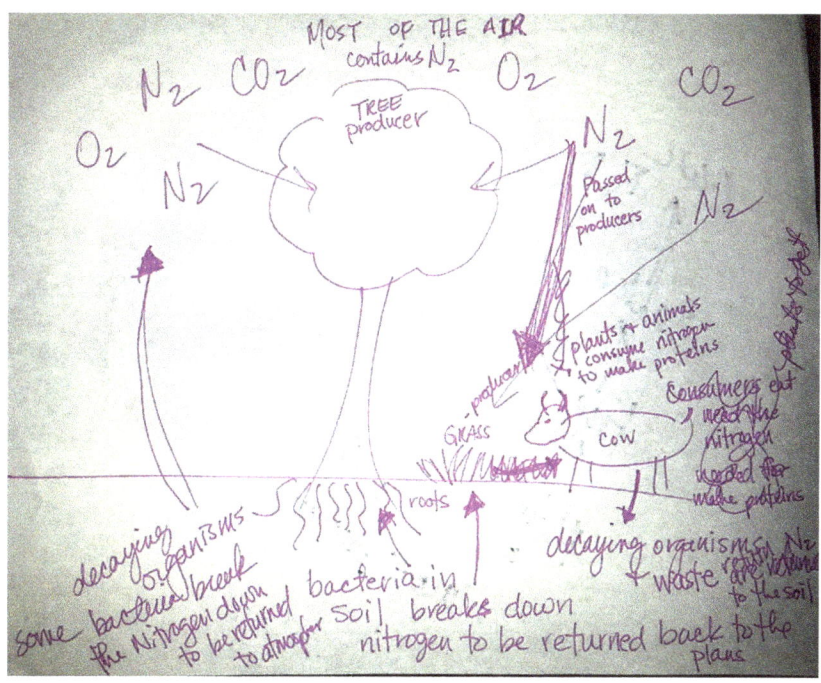

Sky's students' murals grew longer and longer as pages were added to represent conceptualization of each new topic, from food webs and populations to symbiotic relationships. She tasked her students to do their own background research and then synthesize what they had learned in "Know"tations of their own making. All work was completed in class. No one drawing method was privileged over another. Some students used drawing techniques which were highly representational, as depicted in these panels by a GT student:

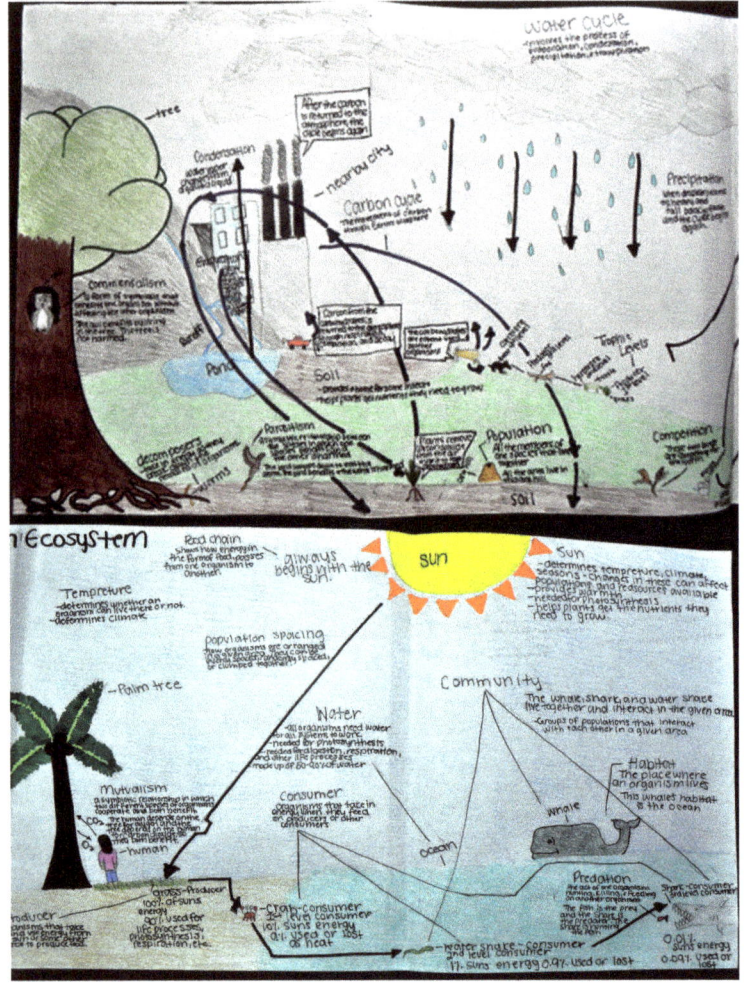

Other students just as lucidly conveyed their understanding of ecological processes with more expressive forms of drawing, as shown here:

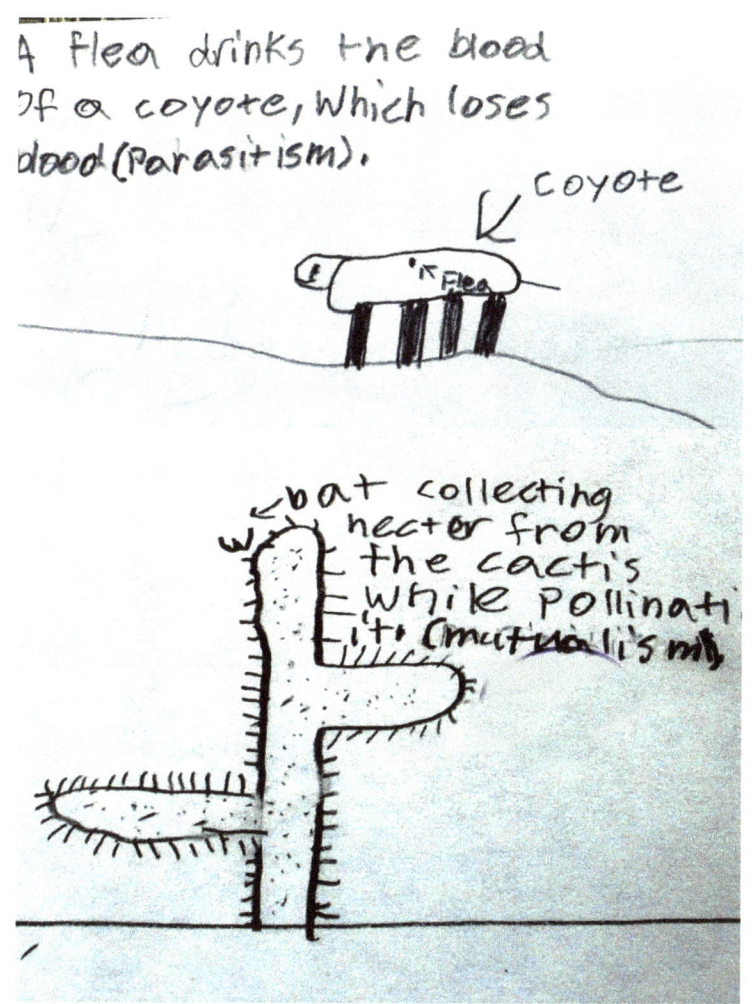

As an integral part of this summative assessment, Sky created a performance rubric, which she gave to both students and parents. Sky later asked parents for feedback on the assignment in general and specifically what they thought about drawing to learn in science. One parent immediately wrote back: *I love it when teachers take a bit of different approach to the curriculum, especially if is something that might be more engaging than the typical memorization of facts.* Another replied: *It was obvious*

this project helped our son study for his test, while a third wrote: *This kind of learning is right up his alley*...involvement rather than lecture. Not all parent comments were positive, however, and it was apparent that the assignment as written was overwhelming for at least two students:

> Unhappy Parent #1: *I think it was too much information for one mural.*
>
> Unhappy Parent # 2: *My son is a create- it –on- the-computer -type –of- kid. I think it would be tough for him and other non-artistic kids if all the projects require a lot of drawing.*

Any negative feedback provided important data points, and will be discussed later in this book. Sky conceded that she could have easily generated four or more separate performance assessments from this one ecology unit. Regardless of any shortcomings either the lesson or its assessment presented during this first trial effort, it was clear from the students' pride in these murals that much learning in science had indeed occurred. Sky's innovative ecology unit was co-performed by teachers and students together—it was sense-based, emergent, connective, empathic, reflexive, inventive, interactive, inclusive, and flexible, with innovations and insights conceived in the moment of epiphanies. Most important, it placed the responsibility for noticing, learning, and creating explanatory models upon the students. Patterns were emerging in our participatory action data. A model of teaching science as aesthetic inquiry was beginning to take shape.

After the mural project was completed, Sky was eager to put her new pedagogy to *The Test*. While my own study was qualitative, my teacher participants had most definitely framed their own action research as an experiment, with teaching science through drawing as the independent variable. Accordingly, Sky now administered the same standardized multiple-choice test on ecology, which she had given on the same the content the previous year. To her delight, there was measurable improvement in all three matched ability groups—students identified as gifted and talented (GT), "Regular" students, and identified special needs students with IEP's. Significantly, the students with special needs or disabilities (those with IEP plans) demonstrated the most test score improvement:

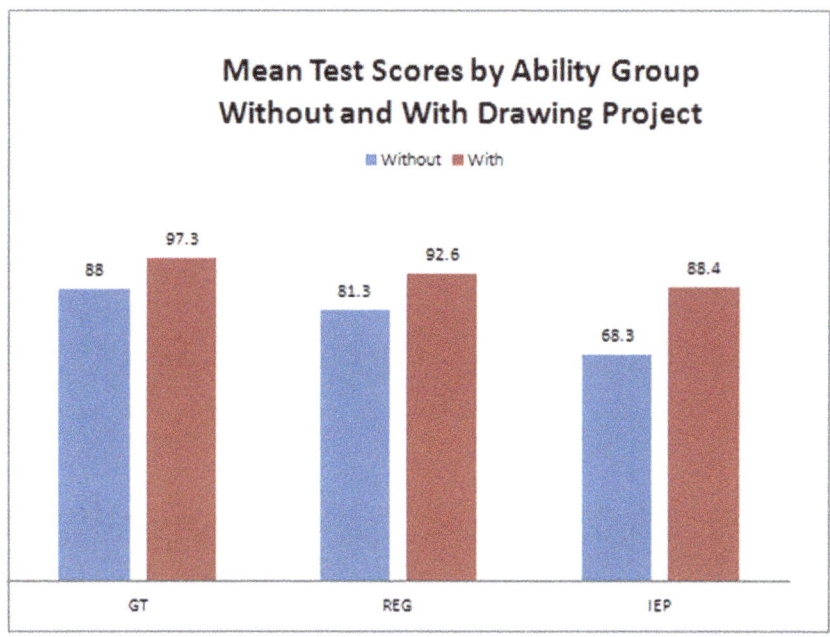

Sky's principal was so impressed with these quantitative findings that he asked her to present them at school-wide faculty meeting in the spring and then again at a fall district wide science teacher workshop. More important, Sky had empowered her students to take responsibility for their own mastery of science content in ways that appealed to them and positively shaped their self-efficacy. The assignments as designed required self-regulation, reasoning and planning—all desirable outcomes of effective assessment practice in science (Black and Wiliam, 1998).

In addition, all ability groups demonstrated improved achievement on a standardized multiple-choice summative assessment. For Sky, there was no turning back now. She had officially become a Science Teacher Who Draws. Would that Eartha had been given such nurturing and support for her research.

LOCKDOWN

last year I had a nature trail

i could use.

    this year,

    i am locked in.

    the principal has

    the KEYS.

i have 168 students,

    some classes with

    38 kids.

    i have a thick file of

    accommodations.

sometimes,

    i enjoy being with animals

    more than people.

    this is one of those years.

Eartha was the very first teacher to enroll in my research study and the first to turn in her ontology statements. No teacher was more eager than she to discover the possibilities of teaching science through drawing, especially if it meant she might reach her students who were extremely low readers:

I WANT TO CONTINUE TO GROW AND IMPROVE MY CRAFT

SO THAT I AM A BETTER TEACHER FOR MY STUDENTS.

WHAT MATTERS TO ME PERSONALLY ARE

MY FAMILY,

MY GOD AND MY CHURCH.

RESPECT FOR OTHERS,

AND

THE TREATMENT OF THE EARTH

AND ALL ITS INHABITANTS,

BOTH SMALL AND LARGE.

Eartha *was* Mother Earth. Her very DNA craved soil and air and rocks and water. She told me she loved "all things sparkly". Her teaching philosophy was rooted in her personal crusade to reverse her students' cultural dis-connection with nature. In our first interview, Eartha spoke with longing about her old principal, the one who had worked *with* her to establish a nature trail and garden on the school's property. She boiled with bile about the new principal, the one who had decided that students would no longer be allowed to have any classes outside. Eartha mourned that her students would not be digging in the dirt that year and that more of them than ever had been put on sedatives that stifled their natural creativity. She eagerly shared, however, that she had begun to incorporate drawing into all of her lessons: *In one assignment, we broke water apart. I asked my students to draw the apparatus and to show that an input of energy was needed. They gave me awesome work and proved through their drawings they were getting it. They struggled with the words because they can't read well, but at least learning was taking place.*

Early in the study, Eartha sent me a "Know"tation she and her students had developed together to visually communicate the concept of "pH":

This explanatory model met all the "Know"tation criteria of having ample white space, a strong, simple central image, visually linked to essential science vocabulary terms and empirical data, and written with typography that was easy to read. Eartha was a natural graphic facilitator of science learning.

During our October Saturday meeting, Eartha told our group about Marcus, who was dyslexic. She said he started out the year missing a lot of days because he was on new meds. But Marcus loved drawing on the table with the dry erase markers (taking a page from Sky's lessons): *It was such a release for him. Before we started using drawing, he was in the back of the room, head down, shut down. Now he's front and center, saying, "I want to do this." Even though his drawing and handwriting skills aren't the best, he's participating. It's like night and day. He passed his microscope test and got a C on a semester exam, when he hadn't passed anything before! This was a standardized test, furnished by the district. You plug in the indicators, and the software generates the test.*

I eagerly anticipated visiting Eartha's classroom to observe her teaching performances. As it turned out, while Eartha and her students were being locked *in*, I was soon to be locked *out*. She had given the informed consent forms for my research study to her principal, who then forwarded them to district office personnel. In November, Eartha received word that no research in district classrooms would be authorized. No outside observers allowed in the classrooms. Period. No explanation was offered. At our December Saturday meeting, all of the other teachers reached out to Eartha to share their wholehearted condolences. Indeed, they all faced obstacles in their teaching jobs, but none had ever been prevented from conducting action research, especially that which stood to benefit students for whom traditional teaching methods were clearly not working. Eartha was beside herself. My IRB guidelines required school and parent consent for me to observe in her classroom. Nevertheless, Eartha continued to attend our group refection sessions. No chains can hold her. She *will* find a way out one day. Of that, I am sure.

6 Create New Vistas

Showing and Telling

At the beginning of this study, Marsh did not care one whit about physical science. He was able to draw almost nothing of significance on his *Back of the Napkin* drawing about force and motion or simple machines, which he described as his "least favorite" area of science to teach. He actually tried to hide his napkin! Marsh cited his own lack of course work in the physical sciences and the fear of being exposed as an "imposter" as the reasons for his previous decisions to leave this topic until the last few days of school each year. He said he had been called out by a middle school science teacher for not being able to read the text (recall he was severely dyslexic) and getting his equations backwards. He had, like a lot of middle school science teachers (myself included) managed to become certified with only the most basic, if any, knowledge of physics. Loving the outdoors, Marsh had earned his masters' degree in natural science. He volunteered to take care of rescued squirrels on Thursdays. He took to teaching *life* science through drawing as if he had been doing it his whole life. For seven months, I watched and waited. Would he take the *Back of the Napkin* challenge and attempt to develop a *physical science* lesson through drawing, too? I said nothing, not wanting to push. Finally, he called me. "Merrie, I've done it! Created a lesson on simple machines! I don't *hate* it anymore! Then, the words started tumbling out so fast, I had to get him to slow down several times.   Here's the playback. You should read it out loud extra fast just to get the feel of this interview:

You know, Merrie,

It just came to me.

Our school is doing

Expeditionary Learning.

Last week,

my principal asked me to

create an EL lesson.

It's project based and collaborative

and gets students to care

about others.

In EL, you also integrate

lots of subjects.

I thought, this is my chance--

I'm going to create a *physics* expedition

right now!

You said to work from a place

of what matters to us.

You know how I love

social studies.

I also used to be a nurse.

Don't forget squirrels,

I added.

How did you know I

put that in,

he asked?

Because you love squirrels, I replied.

I really and truly almost wept with happiness to hear Marsh bear witness to his complete change of mind and heart. We talked about how the Expeditionary Learning approach, (http://elschools.org/), based on the ideas of Kurt Hahn, German educator and founder of Outward Bound, was highly aesthetic in nature, especially in its emphasis on empathy, caring, collaboration, and trust building. I asked Marsh to send me the drawings he had developed for his lessons. He did, and at once, I again realized that I was the one who was the learner here. In one "expedition", Marsh was planning to take his students back in time, to trace the history of simple machines, especially the wheelbarrow. In the process, he would be teaching key science and engineering practices:

His beloved squirrels starred in their own simple machine drawing narratives:

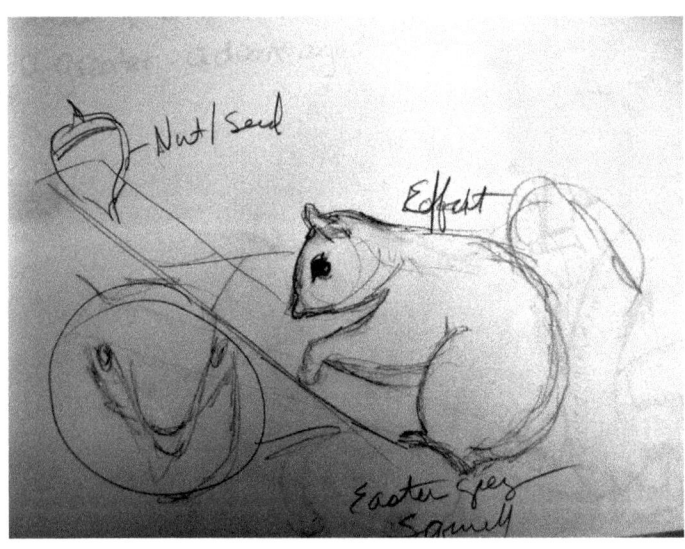

About three weeks into his Expeditionary Learning Project, which he called "Imagination Innovation", Marsh called to say he had a surprise he wanted to share with me. He had created his first ever performative narrative drawing lesson – on simple machines, no less. Could I make the trip? I was out the door the next day.

What follows is a Readers' Theatre script, which, when performed, is meant to re-create what I observed that spring day in Marsh's class (and with which I began this book.) It was the kind of teaching/learning peformance I believe Tharp *et al* (2004) had in mind when they wrote their CREDE (Center for Research on Education, Diversity, and Excellence) report on effective pedagogy, in which they described the following best-practice teacher behaviors:

- Designs instructional activities requiring student collaboration to accomplish a joint product.

- Matches the demands of the joint productive activity to the time available for accomplishing them.

- Arranges classroom seating to accommodate students' individual and group needs to communicate and work jointly.

- Participates with students in joint productive activity.

- Organizes students in a variety of groupings, such as by friendship, mixed academic ability, language, project, or interests, to promote interaction.

- Plans with students how to work in groups and move from one activity to another, such as from large group introduction to small group activity, for clean-up, dismissal, and the like.

- Manages student and teacher access to materials and technology (Teaching Tolerance, n.d)

I can imagine this script being enacted by pre-service educators as a case study and then evaluated to discern which of the above effective teaching behaviors were present in the performance.

I now present...

The Story of Chilli the Mutant Crab and the Bo Jangles Chicken Leg

Researcher: *On the Smartboard in Marsh's classroom is projected the Essential Question of the day: "How can we make machines better? How can we make daily life for elders better?" and the Learning Goal: I can design an innovation to complete a simple task. Along the bottom of a bulletin board at the front of the room are individual small posters with different Habits of Scholarship (H.O.S's) written on each one. Here was the H.O.S. for the day:*

> **I listen to and respect the perspectives of others.**

Since my last visit, there are now posters hung around the room which depict the lives and contributions of different engineers and inventors in history and am pleased to find scientists of color included as well, like Mae Jemison and George Washington Carver:

The lab tables and chairs are pushed to the sides of the classroom, leaving a large space in the center of the room. In this script, transcribed almost verbatim from the audiotape of the teaching performance, I will play the role of the "Researcher".

The bell rings...

Marsh: Good morning, everyone! This morning, I want you to circle up so that we can debrief before you head off for spring break about what we have accomplished so far in our IMAGINATION INNOVATION expedition after one week. Let's talk about what we're getting out of it. Let's get our neurons pumping. What has gone on with our engineering expedition so far?

Student: We've talked about simple machines and built levers together.

Marsh: Right. What else?

Student: We've learned about old people losing their ability to do simple things like we can.

Marsh: Exactly. Remember, the physical therapist called these DBA's or Daily Based Activities?

Student: We also talked about finding ways to make machines better.

Marsh: Awesome! That's the innovation part. What was the name we used for the population of elderly people?

Student: Geriatric!

Marsh: Right. As you're on vacation, I want you to imagine you are real biomedical engineers and that you are going to come up with innovations to make DBA's of geriatric people easier. What are some examples of DBA's:

Student: Tying your shoe.

Marsh: Tying your shoe:

Student: Putting on your shoe.

Marsh: Putting on your shoe.

Student: Eating and grabbing a fork.

Student: Writing.

Student: Taking a bath.

Student: Putting on a jacket.

Marsh: Right! What about zipping the jacket? Buttoning? These are all things we just take for granted but which can be very frustrating and even painful for geriatric patients. Some of the people we will be working with have lost the ability to close their hands.

Researcher: *Marsh holds up a partially closed hand, making it seem crippled.*

Marsh: I want you to come up with ideas for using simple machines to help make our geriatric friends DBA's easier.

Do you remember the mechanical engineer who came to class this week?

Researcher: *Student nod and smile.*

Remember when he told you that *engineers* go through several key steps in their design and problem solving process?

Researcher: *Here, Marsh changes the slide, whose topic is Engineering and Design Steps: Idea, Collaboration, Sketch, Design, and Test.*

Marsh: Remember these, okay, because in just a little while, you are going to become mechanical engineers and solve a problem using these steps.

Marsh: What is collaboration?

Student: Putting ideas together as a team.

Marsh: Indeed! It's one of our Habits of Scholarship!

Marsh: What do you remember about how collaboration works?

Student: You have to be a good listener.

Student: You share your ideas.

Student: It's what we're doing right now.

Marsh: Wow! Excellent. What we're going to do in just a bit is that we are going to start generating ideas that a *special friend* of mine is going to need your mechanical engineering skills to solve.

This will be a practice for something we're going to do after spring break. As part of our expedition, we're actually going to have an Inventors' Fair! You're going to start collaborating on ideas, sketching them out and making inventions to make the DBA's for our geriatric friends easier. You're going to be mechanical engineers yourself!

Students: YEAH!

Researcher: *Marsh's students are about to explode with excitement. One young man pumps his fist in the air.*

Marsh: The senior citizens at the rehab center we're working with are going to give you feedback on your designs. We're going to use all these engineering steps and make these innovations happen!

In the meantime, if you do go back to the senior center, try to feel what your patients are going to be like. For example, many of them have spines that can't support their necks and lower backs properly, so they are slumped in a chair. I want everyone to slump down in your chair with your head down and then see how hard it is to reach straight up.

Student: General murmurings of barely contained excitement.

Researcher: Marsh *puts up another slide: EXPEDITION GOALS for the Imagination Innovation project"*

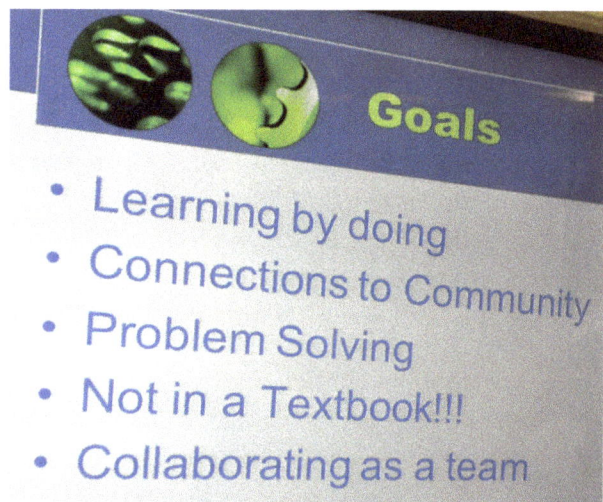

Marsh: Okay, let's review the goals of our expedition. Who wants to tell me what "Learning by Doing" means?

Student: Like, you're not just *talking* about learning. You're actually *doing* something.

Marsh: Yes! It's *active* learning. How have we "Connected to the Community" so far?

Student: People came here from where they worked?

Marsh: Great! How else have we connected with other people as we've learned about engineering and simple machines? Hint: Think about ELA.

Student: We practiced interview skills!

Marsh: Exactly. This will help you make better connections when you do meet our senior citizen friends. For today's activity, we're going to experience more fully what it means when we ask this science question: **"What is the relationship between the Load and the Effort on a lever and how can we create a greater MECHANICAL ADVANTAGE for the lever user?** So far, we've done a couple of labs with simple machines, and we've talked some about the idea of mechanical advantage. What do you remember from our labs with simple machines?

Researcher: *Students struggle with this idea, offer several vague ideas, but clearly don't have any sort of grasp of the science content Paul has intended for them to learn. [Note: Paul has fallen back into his old way of trying to introduce science vocabulary and concepts before actively engaging students in the doing of the knowing. He seems to know this and frowns.]*

Marsh: Hmm. I'm going to change things up. I would like you to get your Ipads and open up the *Educreations* or *Doceri* apps. The Habit of Scholarship I now need you to practice is Listening.

Researcher: *Students scurry to get their Ipads and then expectantly plop themselves down on the floor in a circle in front of the white board.*

Marsh: I think it's time for you to meet my friend. I hope you like him, because he's a little crabby. He's going to appear before you on this paper as I draw him. Are you ready?

Researcher: *All students have their fingers poised over their Ipads, ready to draw. It's clear they've participated in these drawing performances before. I can almost hear their pulses quickening. Several are grinning broadly.*

Marsh: Okay, back to our question about using a machine to create an ADVANTAGE. If something gives you an advantage, what does that mean?

Researcher: *This time, he is deconstructing the question, and starting more with terms they are more likely to already understand.*

Student: It makes your work easier.

Marsh: Exactly. That's what I'm all about! And my friend is going to need your help to make a job easier for him, too!

Student: Is he going to pop out at us?

Marsh: Just draw with me, and you will soon see him! Let's draw a swooping line, like this. And then add two sideways "C"s, and two big circles, like this:

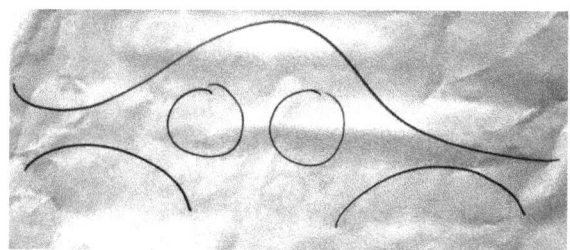

Marsh: The next thing I want you to do is put in what look like two footballs here and close the gap between them with an upside-down "C" shaped line.

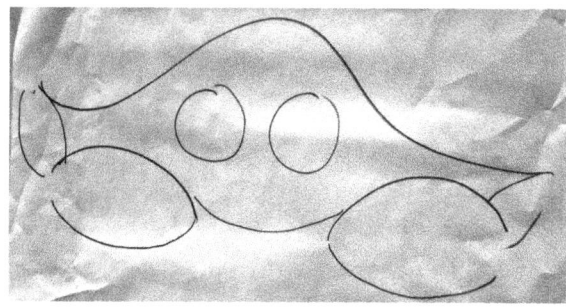

Student: Is he a football player?

Marsh: I guess he could be!

Marsh: Now, let's add some upside-down "V's" along the top and put some squiggles inside those football shapes:

Did I tell you this guy is my best friend.

Researcher: I thought I was your best friend!

Marsh: Oh, you are, but don't you remember this fellow?

Researcher: *(playing along):* Oh, that's right. I do remember now.

Marsh: Yeah, don't you remember how crabby he was?

Student: several simultaneously shout, "He's a crab!"

Researcher: *laughter all around...*

Marsh: I can't really introduce you to him yet, because he's not really together yet. Just a little bit longer. We need to put in his mouth and eyes, and some little eyebrow looking things.

Researcher: *Little crabs are appearing on Ipads all over the room.*

Marsh: And yes, he is a crab, but he is not just any crab. He is a MUTANT crab. He looks like a cross between an Atlantic Blue Crab and a ghost crab. His favorite pastime, besides eating food you leave behind on the beach, is just to chill out. That's why his pals named him Chilli. Say hello to Chilli!

Students: Hi, Chilli!

Marsh: All crabs are scavengers, which means they'll eat pretty much anything, but Chill's favorite food in the whole world is Bo Jangles chicken. Last week, the beach maintenance crew put a trash can right next to his den up near the sand dunes. They also planted a palm tree:

Chilli suddenly had the best spot on the beach! The lovely smells of people's discarded food attracted bossy sea gulls, however. They're also scavengers. During the day, Chili stayed hunkered down in his den, BUT...

Just after sunset, however, when most of the people left and most of the birds went to rest, the beach belonged to Chilli'! Last evening, just after the maintenance crew had emptied his trashcan (always a sad moment for Chilli), an eleven year old girl, leaving the beach with her parents, dropped half of her Bo Jangles drumstick into his trashcan.

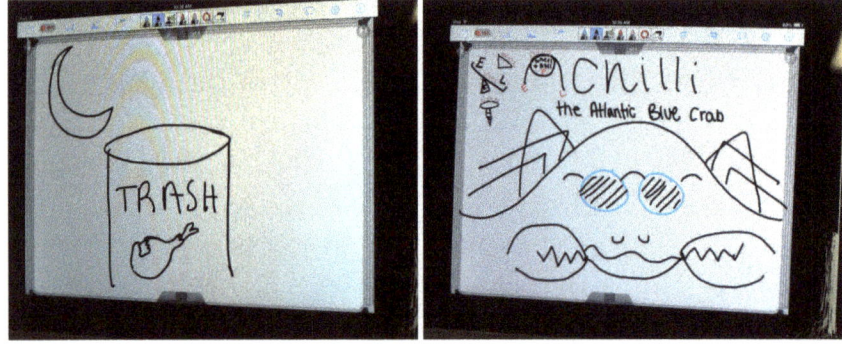

Chilli's beady eyes almost popped off their stalks! *Thunk!* Chilli heard the Bo Jangles drumstick hit the bottom of the can. He fretted and blew spit bubbles. How in the world could he ever get that chicken leg? He was, after all, just a little mutant crab. Oh, gee, it smelled wonderful!

Marsh: Class, Chilli needs our engineering help! We need to help him get that drumstick out of the trash can! We've got to come up with some ideas.

Researcher: *Hands shoot up, and Paul calls on one student.*

Student: Well, he could use a lever.

Marsh: A lever! That's brilliant. Will you come up and draw what you mean?

Researcher: *The student draws a stick wedged at an angle at the bottom of the trash can. She then draws a rock under the stick.*

Marsh: Who can help me figure out where the "E.L.F.s" are, the Effort, Load, and Fulcrum? I'm having a hard time remembering. What should we label first?

Students: The fulcrum!

Marsh: Where is it on this drawing? Will you come up and write an "F" over the fulcrum?

Researcher*: A student comes up and draws the "F" for fulcrum, on the rock.*

Marsh: Where's the Load?"

Students: The trashcan's the load! They reach out for the marker.

Researcher: *Paul hands one to a student, who writes an "L" on the trashcan.*

Student: And Chilli makes the Effort! Can I draw it?

Marsh: Of course, here you go!

Researcher: *He hands the student the marker.*

Marsh: Okay, this is great! How can we use S.T.E.M. language to describe what Chillli needs us to do?

Student: We're making his work easier.

Researcher: *Paul writes "Make work easier" next to the trashcan and students do the same. Their visual notes are evolving with the lesson, as they discuss all the various simple machines they might use to help Chilli.*

Marsh: Awesome! Technology is all about using Science knowledge to make jobs easier. What about the science behind the *lever* you drew? Or the pulley and inclined plane ideas you suggested? What *science* language would you use to describe how those simple machines would help Chilli get that drumstick?

Student: Oh! Oh! We're giving him a mechanical advantage!

Marsh: Is that right? Does everyone agree? Does a machine that gives you a mechanical advantage make the work of getting that drumstick out of the big metal trash can easier? Let's write this on our visual notes together.

Researcher: *Paul draws an arrow to the right of the words "makes work easier" and then writes "mechanical advantage."*

Marsh: Wow. You guys are amazing! So far, we've completed the first two steps in the engineering process – We've come up with some IDEAS about solving Chilli's problem and we've COLLABORATED on ways in which we might help him get that drumstick using simple machines. Now, it's time for my favorite parts – the SKETCHING and the DESIGN of your possible solutions. For this part of the lesson, I'm going to put you into teams of 3 to sketch out and agree on the design of the simple machines you would build to help Chilli get to that drumstick!

But let's remember. Chili can't just go to Lowe's and buy materials. He has to make use of what he may find on the beach. Before we start, let's make a quick list of what materials he might find.

Student: People leave all sorts of things behind them on the beach.

Student: Yeah, I found an entire beach umbrella once.

Student: And a pair of shoes.

Student: What about things that wash up on the beach?

Marsh: Awesome. What kinds of things might wash up?

Student: Fishing things- like fishing line or even reels.

Student: Or pieces of wood!

Student: Or shells!

Researcher: *Marsh writes all these suggestions as they are offered.*

Student: I've never been on a beach.

Marsh: Okay, we're going to bring you there in your imagination today. Are you with me?

Researcher: *Student nods.*

Marsh: Okay, I think you have the idea. There's just one more part of this assignment. I want you to create a *story* that tells how Chilli goes about making the simple machines to get that drumstick. Remember that you can combine them together. When your group agrees on the design, elect an official graphic artist and narrator. Press record on your drawing app and have the graphic artist draw while the narrator explains what Chilli is doing. Then, email the final production piece to me for your formative assessment. We'll meet back in 15 minutes and play each group's video. Maureen, and Ely, I want you to be timekeepers and remind each group how much time they have left.

Researcher: *This has been a phenomenal lesson. Marsh has invited students to co-construct their "knowing" in ways that were highly enjoyable, meaningful, and relevant, not to mention standards-based. Every single student was fully engaged in the process of engineering a solution to Chilli's problem. Now, he was giving them the freedom to use their imaginations and creativity to apply their science knowledge to develop a* *technological and engineering solutions using simple machines. How I hope the play I have written to re-present the experience will inspire others as I have been.*

###

VOICE WORDLES

At about seven months into this research study, I asked my teacher participants if they would be willing to take a few moments to ask their students reflect on how, if at all, learning science through drawing had affected them. I created the following reflection questions: 1) How does drawing in science make you feel? 2) Has drawing in science helped you better learn and remember the material? If so, would you describe what exactly has improved? ? 3) Has drawing in science had any kind of negative effect on your ability to learn science or your willingness to learn it? If so, will you describe how or why this has been the case? I received over 300 completed student reflections. I coded them in terms of the 1) feelings/emotions communicated, 2) the action verbs which were chosen, and 3) the perceived learning outcomes, making note of word use frequencies. I then created word clouds, using the *Wordle* app available at http://www.wordle.net/create. Like other word cloud generators, the size of the font increases with the frequency of times the word is entered, which allowed me to visually scale students' responses. The larger the font, the "louder" the voiced opinion.

From the emotion codes of the *positive* responses (97% of the total) arose what I have called *The Feelful Wordle*.

Not all students had positive feelings about the experience of drawing in science, however, even though no student was required to draw. Nine out of the 300 who submitted reflections did *not* feel comfortable with drawing in science. Significantly, these were mostly students identified as gifted and talented, who believed that their lack of drawing talent would negatively impact their grade. Also, all but one of the negative responses came from students of a teacher who required students to create multiple projects for summative assessment in which drawing was required. For these nine (out of 300) students, for whom drawing in science was definitely *not* fun, I generated the following *Stress Wordle*.

It was quite apparent from these very honest reflections that student comfort levels with drawing need to be addressed from the very beginning. Even though all teacher participants made a point of communicating that artistic talent would not be factored into grading assessments, there were still some students whose perceived drawing incompetence generated significant anxiety when they knew they were being graded for projects which required drawing. How and when drawing should be used in summative assessments is a matter which warrants significantly more dialogue. The action verbs submitted by students who submitted positive responses (97%) resulted in this *Action Wordle*:

Finally, from the outcomes of the 97% who communicated positive feelings, I generated the *What's Happening Here Wordle*:

Synthesize

7 *Science, Philosophy, and Aesthetics*

Tangibles and Intangibles

One of the most unnerving experiences of my undergraduate study was being tasked with discovering the identity of an "unknown". Each of us analytic chemistry students was given a thin-walled glass tube inside of which was a mixture of compounds, whose identities would become "known" to us if we could somehow get the nuclear magnetic resonance (NMR) spectrophotometer deep in the dark bowels of the science building to work properly. We were allowed limited time to "interact" with the gray metal beast. An enormously high percentage of our grade depended on our getting the right answer. To my horror, when it was my turn to sit at the console of the NMR, my normally agile fingers became spastic and I broke the tube. I was given another and several agonizing weeks later, successfully identified my "unknown", but I never took another chemistry course again. With the shattering of that tube of glass (and my confidence), I entered a dialectic with myself that has continued to this day. The previous semester, I was so enthralled with organic chemistry that I could be found any given night happily drawing compounds in various iterations on the blackboards of the classroom building. Spread out on the desk were ball-and-stick models, ready to be turned into the isomers of my choice. I was mesmerized by the concept of chirality. But now, I was in complete crisis. I did not cut it in the lab. Did this mean I was *not* a scientist? Could not *be* a scientist?

I was at that same time also taking my first ever philosophy class, an experience which forever shaped my ways of thinking, being, and doing. In Dr. Edward's brightly lit classroom, far removed from the ominous NMR that threatened my GPA, we students engaged in lively debates in which there was plenty of room for deduction and induction, for quantitative and qualitative, for physics and metaphysics. In philosophy, there was an entirely different type of *analysis* going on: *What is real? What knowledge is of most worth? Who gets to decide?* I have devoted most of my career trying to resolve the perceived contradictions between the nature of science and art. In developing my own philosophy, I have been deeply influenced by my reading of Hegel. To me, Hegel's dialectics were developmental, even evolutionary in their approach. His writings challenged me to develop the habit of preserving what is best about even the most apparently incompatible positions to arrive at a new location, which represents a new, higher, more hopeful and even synergistic solution.

I have come to understand that scientists, artists, and philosophers all engage in the practice of *synthesis*, the act of making something new that did not exist before. With few exceptions, synthesis proceeds from the act of analysis, which asks the forensic questions, *What is there? What is happening?...*

THE TREE OF "KNOW"LEDGE

Researcher: So, what has happened here?

You all signed on

with the hope that

it might help

	your struggling students
	succeed (*long pause*).
	Right?
Maya:	The way I was teaching before was
	not working (*sighing*).
	But I was afraid to do this research.
Researcher:	Why was that?
Maya:	I didn't draw well (*sighing louder*).
Researcher:	But you cared enough
	to try.
	All of you did.
	And caring opens clearings in which
	learning can happen.
	Each of you created different
	drawing innovations and thus,
	different clearings and unique ART.
	But there were common themes throughout

	That made teaching science through drawing work.
	But...what did you *do* exactly?
Marsh:	We drew our brains out.
Researcher:	And then, what?
Crystal:	I deepened my own science content knowledge.
Sky:	I stopped talking so much and turned my students' attention toward noticing things in the natural world, beginning with being still.
Marsh:	I drew because I have so many poor readers in my class, and because they come alive when we use drawing to learn science. I'm also dyslexic myself, and I know what it's like to struggle with science vocabulary.
Maya:	I drew with my students because it challenged them to show me what they knew and it helped them review for tests.

	Even my dyslexic students started passing.
Researcher:	So, you are telling me the research worked, right?
All but Eartha:	Right!
Researcher:	*(Looking thoughtful, then picking up a pencil and starting to draw in her sketchbook)*
	Imagine this line is the soil, and that during our research, we grew a *garden* of students. Really *healthy* students. Let's draw some lines below the soil to represent the roots in our garden (which you never see). Now, let's connect those roots to some some stems (which are clearly visible). Let's imagine that each stem represents a different kind of student, each producing different shaped leaves. Let's call our model,

WHY IS THIS WORKING?

What tangible growth factors could you SEE (above ground)?

Which intangible factors would be INVISIBLE (below ground)?

Now, your turn...

Marsh: We took risks. *(Picks up the pencil.)*

Crystal: We asked our students what they were feeling. *(Picks up the pencil.)*

Maya, Sky, Marsh: We had support from our
(in unison) administrators and parents. *(Pick up the pencil.)*

Eartha: *(frowns).*

Narrator: Slowly,

a success story,

signified

by simple lines, shapes, and words,

emerged on the paper.

There was no mistaking

its meaning.

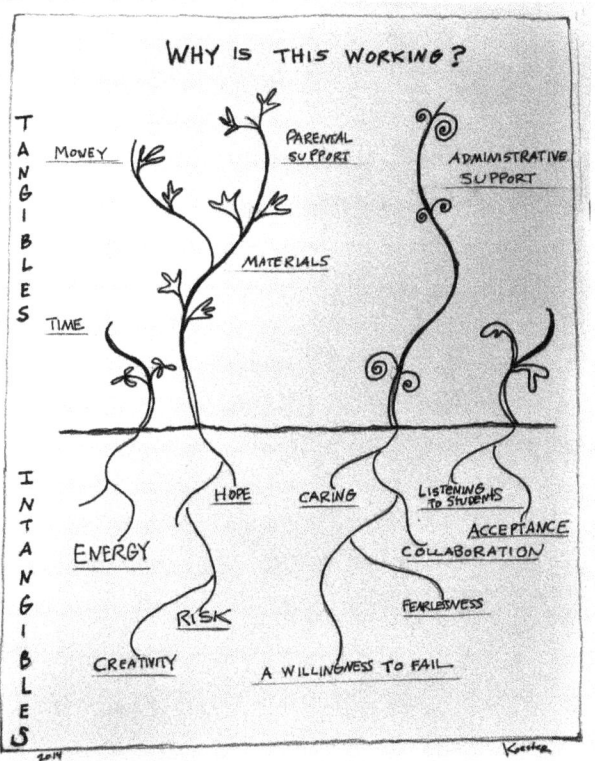

Researcher: Paul Klee wrote:

I don't want to

render

the human being as he is,

but rather in the way

he might me.

This, too, is the reason we all teach.

Both scientists and artists create models to re-present their interpretations or explanations of phenomena. By their very nature, models are theoretical constructions and as such, always subject to revision. Windshitl et al. (2008) have explained that *scientific* models share five key characteristics: They are 1) testable, 2) revisable, 3) explanatory, 4) conjectural, and 5) generative. The relative "value" of a scientific model is determined by its ability to predict measurable outcomes, explain past events, or inform future research. Lack of scientific precision (or working from a faulty model) can mean the difference between living and dying, a rocket not reaching its destination, national security being compromised, or the approval of a medicine with unforeseen side effects. I will argue that the "best" *science* models arise from sustained sense-based, thus *aesthetic,* acts of *noticing* what there is of significance about the subject in question. I believe that what Greene refers to as noticing is about finding *patterns* in sense-based data. McKim (1980) has stated that "pattern-seeking is the first step in all perception...[and that] the process of organizing stimuli into simpler groups is almost involuntary...whether a person is a painter grouping visual elements in a painting, a scientist classifying observations, or students grouping their notes into outline form" (p. 65). *Conjectural* thinking and the *generation* of new questions are *creative,* hence *artistic* cognitive acts that arise from the communicative, connecting, aesthetic dimensions of science. Because they are *meta*physical, these creative thinking performances cannot be measured but are nevertheless essential components of the practice of developing exemplary science models.) Finally, because creating science models involve "acts of making that require intellectual judgment", they are, by definition, *art* projects, too. Scientists, however, have rarely acknowledged the connection. Botanical *illustrations* rarely portray the root systems of plants (Stafleu, 1968) either, but we all know they're there. I have facilitated hundreds of workshops on teaching science through the arts. "But where is the *Science*?" is a question I often field. "What we're doing is *just* art." (translate *play*). I have learned to make my sighs inaudible, but know I am grimacing loudly. The *red* is an integral part of the *nature* of science. Without red, green cannot sing. Without air, there is no sound. Without art, there is no engineering or invention. Making science-based art (whether that takes the form of embodied role play, a drawing, a simulation, a poem, musical narrative, animation of a science process, etc.) creates a medium for shared understanding of science matters.

Truly, when art is *othered* in the science (or any other) classroom, *everyone* loses. As Eisner declared again and again, not everything that matters can be measured. During the performance of this research, we could all see that something of substantial significance had occurred. I sought to create an analogic process model (a drawing, of course), which included both the tangible and intangible factors which had contributed to the success of this aesthetic approach to science education—an artifact that at once re-presented both and analysis and synthesis of the data. It is the "garden" drawing featured in the poetic transcription of my team's last group reflection session. The inspiration for this model came from my reading and appreciation of the work of John Berger and Paul Klee as well as my own love of gardening:

Berger, art critic, painter, poet and novelist, wrote, "We who draw do so not only to make something visible to others, but also to accompany something invisible to its incalculable destination" (2011, p. 9). This metaphor of art as a transport medium between invisible and visible also played a central role in Klee's art and as well as his philosophy. For Klee, the artist was like the trunk of a tree, "gathering and conducting whatever it is that comes to him from the depths." (in Sallis, 2012, p. 10). His 1935 drawing, *Little Tree* (*Bäumchen*), has the liveliness of a musical conductor, directing the magical symphony that is photosynthesis. Klee, the artist, clearly "got" xylem and phloem. I am an avid botanist and have spent many hours drawing plants from both microscopic and macroscopic perspectives. From nature drawing, I have acquired deep, personal, *artistic* knowledge and appreciation which has informed and vastly improved my teaching about life *science*.

Plant collectors prize the flowers as much as policy makers prize high test scores. And yet, no plant with the potential to produce flowers (and ultimately fruit) can do so without first developing robust root systems. The caring, artistic teacher recognizes the growth and survival potential of *all* students, knowing that they, like mangroves, even if they are stuck with poor, hypoxic soil, can, with the right growth factors, form a network of adventitious prop roots to anchor them as well as upward growing "breathing roots", or pneumatophores, which act like snorkels (Landau, n.d.). Caring stabilizes students and triggers the meristem of their potential to build root tips that probe soil they might not otherwise have explored—at least not in school. The drawing artifacts from this study (and the higher test scores) re-present the fruits produced by learners (students, teachers, and one teacher educator) experiencing science as aesthetic inquiry. By grafting aesthetics onto the stems of their pedagogy, teachers increase the chance that diverse and meaningful learning outcomes can be successfully propagated.

Consider the conceptual model of "success" jointly created by my research team. Not surprisingly, the *intangible* (below ground) "growth factors" were all *aesthetic* in nature. To these teachers, no one *tangible* factor was as significant as administrative support. Without it, they agreed their vigorous growth as professional, autonomous educators would be greatly compromised, if not impossible. Eartha didn't have *any* administrative support; so, for awhile, she went "underground". We nurtured her as best we could, but it wasn't enough. Without light, virtually all plants die. I sincerely hope Eartha will seek to be transplanted into more fertile soil.

FAILURE TO THRIVE

By its very existence, darkness defines what light is not. The challenges which prevented Eartha in particular from achieving her objectives, and which transformed her initial exuberance into despair weighed on all of us throughout the study. I knew that as part of our final shared analysis, we needed to address both her pain and perceived failure. Accordingly, I also presented to the group another sketch, which I had entitled FAILURE TO THRIVE. Again, I had drawn a "soil" line. On top of the soil surface, I had drawn a stunted looking, fanged, two-headed creature, attached by an umbilicus to an ominous, swollen looking "taproot" structure. To me, it resembled "Dr. Doolittle's" *pushmi-pullyu*, a conflicted creature that can't go anywhere because both heads pull in opposite directions. Similarly, as long as she was being pulled backwards by her administration, Eartha could not move foreward with her own action research. I asked teacher participants to identify those "root" causes which they believed could contribute to exactly *nothing* happening, i.e., the null hypothesis becoming a reality. Onto the drawing, the following perceived *limiting factors* appeared: *burn-out, being left on my own, no collaboration, apathy, poor salary, emphasis on conformity by administrators, lack of support,* and *tunnel-vision (applying only one way of teaching or research),* and *linear thinking.* These observations were technically measurable in sighs. How different, and yet equally meaningful was this model was from the last!

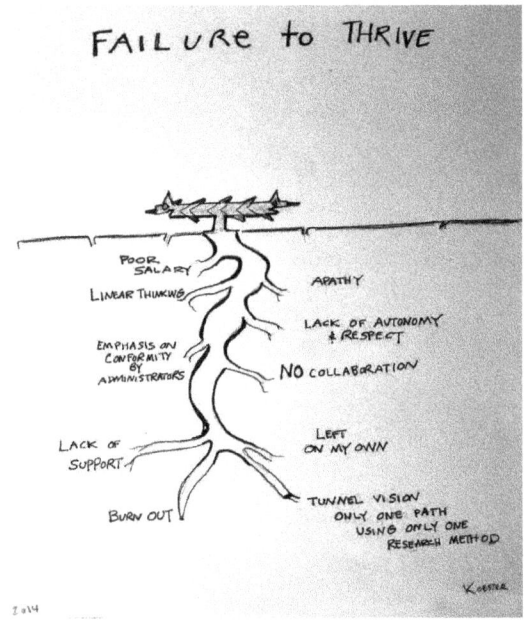

There is a reason the metaphor of "opening spaces" has become so prevalent in education research literature. Closed systems, like terrariums and aquariums, all too frequently shut down because of the accumulation of toxic waste and/or the loss of oxygen. Because Eartha's "life support system" was cut off, her research died shortly after germination.

ENERGY TRANSFER AND *UMWELT*

Back at home, I considered the two jointly produced models we had created during that final reflection session. In conceptualizing these process models, we had evaluated as objectively as possible those factors which teachers believed either contributed to or stood in the way of "successfully" implementing this new paradigm of science education. I believed the plant-form models truthfully and accurately re-presented the "why it this happening" patterns in the data; however, they were too simplistic to signify the "how's". For starters, human beings are not plants. They cannot make their own food, and they are far more tender. They have complex emotions and beliefs which drive their decision-making and behaviors and render them vulnerable. Hurt feelings leave deep and abiding scars which can be life altering until and unless they are addressed.

I next considered whether a more systemic ecological model might more closely approximate what was happening. The German language has a word, *Umwelt*, which Kull (1998) a semiotician, has described as "all the *meaningful* aspects of the world for a particular organism" (no pagination, my emphasis). We humans are not likely to invite something into our *Umwelten* (the plural form) unless we care about it. German biologist and ethnologist Jakob von Uexküll reasoned that the more complex an animal, the more complicated its *Umwelt* and the more difficult it is to determine what makes tick (Chien, 2006). Like all living systems, the classroom is characterized by a reliance on feedback, signals, and response to change in the environment. The effective teacher discovers ways of successfully pinging signals into students' *Umwelten* so that wide open communication networks are established. We had determined that *aesthetic* signals *pinged* with practically all students. Indeed, communication was taking place, and at a high rate, too. But what was driving and sustaining *semiosis* in such a way that students were inviting science into their *Umwelten,* instead of blocking it out, as they had before?

I re-played in my mind the many teaching/learning scenes I had witnessed. I studied every drawing artifact, and re-read every transcribed interview. I finally found what I was seeking in a "Know"tation I had created (fifteen months earlier) during a sixth grade inquiry lab, presented by S.T.E.M. educator, Becky Cornwell, and her colleague, physicist Jeff Wilson, at a summer science institute facilitated by USC professor, Christine Lotter. I laughed out loud at the irony of what was now playing out. My own teaching *Umwelt* (comfort zone) had always been in the life sciences first, then chemistry, then geology. To this day, I do not feel qualified to adequately teach physics. But here was a physics drawing which fit the story of our research (and explained how we got the results we did) better than any process model I had considered.

My "Know"tation featured a beaker of boiling water (on a hot plate) with bits of paper, rice, and other flotsam steadily rising and falling in a circular vortex:

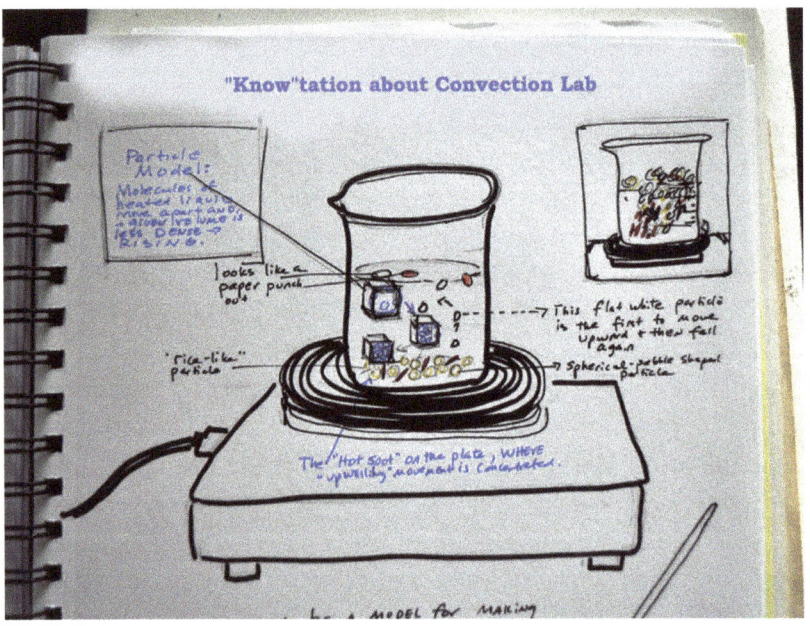

I saw it all quite clearly now. Students were being transformed through a *convective* process. Previously bored, struggling or failing students were being swept up by caring, knowledgeable teachers, who were sources of renewable, sustained *aesthetic energy*. Aesthetic energy was being transformed into rising (and measurable) levels of tangible work and smiling faces. I realized that *felt* convection (a *sense* of being swept up) is what must happen for something "other" or "strange" (in this case, science) to be invited into a student's epistemological and axiological *Umwelt*. I picked up my drawing pencils and finally produced the "how is this happening" model I was seeking:

Anyone walking into such a classroom could see that teaching science as aesthetic inquiry is all about *human* energy transfer, a process which is maximized when the arts are employed as information conductors. Bresler (2006) drew similar conclusions when she observed that "the aesthetic inquirer is at the vortex of the movement, actively seeking connections" (in Latta and Baer, p. 94). Art connects, communicates, and breaks down barriers. The red is always there! We have shown that teaching science through drawing in ways that are aesthetic can transform students who have struggled or even failed to achieve in science into those who eagerly and enthusiastically embrace what there is to be discovered.

I believe that that teaching science as aesthetic inquiry can help achieve the goal of science literacy for all students and that this paradigm should not only be included in science teacher education programs but become the subject of further participatory action research and professional development efforts. The new S.T.E.A.M. (or S.T.E.M. and the Arts) movement is rapidly gaining momentum, and miles of extra track, especially in my great state of South Carolina, are being laid down. Shulman (1987) concluded that that the more we learn about teaching, the more "we will come to recognize new categories of performance and understanding that are characteristics of good teachers, and will have to reconsider and redefine other domains" (p. 13). Lederman's (1992, 2007) reviews of the research on the teaching of the Nature of Science ultimately concluded that the "instructional approach, style, rapport, and personality of the teacher are all important variables in effective science teaching" (2007, p. 845).

I believe that unless the teaching "variables" Lederman described crossover into the domain of aesthetics, there will be a very high chance that what really matters about science will be lost in translation. An aesthetic orientation to teaching implies that teachers can't just script and *direct* student learning, they must, *with* students, jointly *produce* learning *performances*, at the end of which they must also assess student emotions and *feelings* about what just happened (or didn't). Indeed, feedback is crucial, but it can't just be about whether students have memorized the content. Conquergood (1998), has stated that the performance paradigm privileges 'an experiential, participatory epistemology' (p. 27), and believed that only through shared emotional experiences can a performance achieve transformation. Kids want to be part of something where they can be "all in", whether it's the *Algebra Project*, chess club, school orchestra, or robotics team. In each of these challenging, participatory, interactive communities, both teaching and learning are raised to the level of artistic *performances, appreciated* for their aesthetic *qualities* as well as measurable academic outcomes they may produce. It's now time to address the question, "How does one go about *performing* science as aesthetic inquiry?"

8 A Model of Science as Aesthetic Inquiry

What "checklist" of quality performance behaviors might be followed by a teacher who seeks to break away from the traditional transmission mode of teaching science to a pedagogy that is aesthetic? Kelehear (2008) described teaching as a *performance*, which as such, requires the teacher to be fully present and keenly aware of the efficacy of her performance. He wrote a highly informative analysis of ways in which school instructional leaders might address and appraise the artistic dimensions of a teaching performance, positing in his study that if teacher leaders were taught specific reflective methods of art criticism, there would be more likely to evolve in their schools "an atmosphere that supports inquiry, artistic expression, and exploration" (2008, p. 243). He stated that in order "to see teaching as performance, we must be capable of seeing as an artist, and proposed that teacher leaders adopt what Eisner (1972, 1976) has called the *connoisseurship* approach, a particular kind of qualitative inquiry which he described as the art of appreciating the *qualities* of a thing or experience, be it a fine wine or a classroom teaching performance.

Adopting an aesthetic approach to teaching, while emergent and reflexive, is not a haphazard or unscientific one at all, especially when one considers that the paradigm invites observation and experimentation and foregrounds the *creative* nature of science and engineering. The process, as I see it, begins with the mending of mindsets. *Artistic* science teaching as far less about *telling* what is right and wrong than it is about *revealing* what is significant and meaningful so that students can interact with the content and each other. Performing artists spend substantial amounts of time getting "into character". They have engaged in deep study of the life, history, emotional and physical attributes of their characters. Likewise, the science teacher who would deliver a confident teaching performance should acquire deep subject matter content knowledge. Formative assessment by teachers of themselves is as important as that conducted to evaluate student learning. We have shown that all you need is a napkin and a few markers to conduct a brief, but honest diagnostic. What you draw with your students, however, cannot contain symbology (either word or image) that is unfamiliar to the lived cultural experience of your students. Such a "map" will remain indecipherable and perpetuate the separation between teacher and student.

CHANGING TEACHER BELIEFS AND PERSPECTIVES

What kinds of teachers might be willing to experiment with teaching science as aesthetic inquiry, knowing they must engage in cross-training in the arts? How much time is required for the training to "take" and be converted into practice? The answers are a largely a function of teachers' existing beliefs, priorities, epistemic orientations, and willingness to make themselves vulnerable. Quite often, I am met with significant resistance when I suggest that science teachers vary the brushstrokes and/or palettes of their pedagogy. Many fear being exposed for not knowing their content, and prefer the safety of lectures and canned power points. Lacking deep knowledge, however, many ask mostly recall-level questions from prescribed curricula. If they suffer from drawing anxiety, some science teachers may ask *students* to draw more, but do not do so themselves. This is a start at least, for drawings quickly make visible what is being learned (or not). The Frankel (2010) and Tytler *et al* (2013) studies cited in Chapter 2 confirm the value of drawing as a powerful form of formative assessment in science.

In Chapter 4, the concept of aesthetic inquiry as a curriculum of *care* was explored. Peter London's (1989) exposition on how to caringly lead someone through an unfamiliar creative encounter until they can (through self-discovery) form their own personal meaning of art is highly congruent with literature on conceptual change and the efficacy of professional development experiences. Navigating that "period of disequilibrium" (London, 1989, p. 79) can be very tricky. The goal is to afford the science teacher with repertoires for teaching practice which are securely aligned with the new understandings they are constructing (Huberman, 1995) until they can ride the bicycle without falling.

Van Driel *et al.* (2001) reinforce the importance of identifying teachers' deeply held beliefs at the start of a reform project (p. 137). Like Van Driel, I don't think teacher educators can overestimate the importance of establishing what defines a given teacher learner's *Umwelt*. At the very beginning of our research study, I asked participating teachers to submit a "Teacher Ontology Statement" as well as an essay titled "What Matters to Me," both of which served as the focus of our first Skype interviews. These were "free writes", with no parameters or guidelines other than the question. In the ontology statements, I asked teachers to reflect on ways in which they were taught science and how they perceived a "successful" teaching/learning experience looking. In the axiological reflection (What Matters to Me), I was able to get a sense of the extent to which they could find value in grafting an aesthetic approach onto their existing epistemologies. Crystal's first "soliloquy" (Chapter 1) was a poetic transcription of both her ontology and her axiology. During the crucial first interviews, I invited teachers to tell their stories and to describe what they hoped would happen as a result of participating in this collaborative research. In addition, I asked them to identify specific students whom they believed would most benefit most from learning science through drawing. I asked them to document with writing and drawings both theirs and their students' transformations during the research experience. In all interviews, individual and group, participant researchers were encouraged to give voice to their personal struggles to effect meaning-making in the science classroom. We paid very close attention to the significant stress induced when required to teach a subject for which one's own pedagogical content knowledge is far from deep. Participant researchers were encouraged to identify what worked and what did *not* to effect maximal meaning making in their science classes. There was a high degree of commiseration and rapport going on. During our Saturday sessions, we worked together to brainstorm each other's explanatory drawings. We drew and re-drew until we agreed that the visual explanatory model would successfully communicate the *gestalt* of a science concept or phenomenon in ways that were meaningful and relevant to a middle school student, regardless of reading ability.

We acknowledged that the most significant obstacle to embracing the teaching of science through drawing is that many science teachers fully *believe* they either *can't* draw or their students will somehow think less of them if they draw like nine year olds. We agreed that some teachers will take more "loosening up" than others. This was certainly the case with Maya, who at first resisted drawing *with* her students, but who discovered the camaraderie that can evolve when a teacher dares to be human and fallible, too. All the teachers in this study discovered that taking an aesthetic approach to teaching science created opportunities for their own growth as well as their students.

RESEARCH, TEACHER PRACTICE, AND STUDENT LEARNING

This book is very much about the influence of research and professional development experiences on teacher practice. In their literature review on this vital subject, Loucks-Horsley and Matsumoto (1999) documented the importance of teachers' self-assessment and collaboration: "Teacher learning is enhanced by interactions that encourage them to articulate their views, challenge those of others, and come to better understanding as a community" (p.261). Repeatedly throughout our time together, teacher researchers talked about how it important it was for them to be given the time to "vent", share innovations, collaboratively problem-solve, reflect, and above all, be validated as the professionals they were. The Loucks-Horsley/ Matsumoto review also called for more studies which looked at the effect of professional development on *student* learning. Indeed, our team did not truly realize what we had accomplished together until we asked the students how they *felt* about learning science through drawing:

> *It makes me feel great, because it's the only class you get to draw, besides art, with imagination.*
>
> *It makes me feel like I want to come to science every day.*
>
> *When I close my eyes, I can visualize my notes.*
>
> *It makes me feel that I have more freedom by drawing the way I want to.*
>
> *t makes me feel more confident about tests and quizzes.*
>
> *It makes me feel like I'm getting help.*
>
> *Drawing in science has made me better in getting things done right.*

> *Drawing in science makes me feel happy and ready to learn anything.*
>
> *It helps me study. It inbeads it in my brain.*
>
> *It helps me picture things. My grades have improved by a great amount.*
>
> *My addidute is much more joyful.*

Significantly, many of the student responses above were provided by ELL students, for whom the visual representation of science content by teachers and their peers allowed them for the first time *not* to be handicapped by their lack of fluency with the English language. By creating jointly produced explanatory drawings of Big Ideas in science, the teacher participants in this study made possible the active participation of their ELL students in the learning performance. By challenging all students to create "Know"tations of their emerging knowledge as science inquiry and problem solving progressed, teacher participants could formatively assess each student's understanding in ways that included each one in the collective effort. Clearly, students' lives had been "brightened and enriched" by being afforded aesthetic capital (Adjibolosoo, 1995). Their ways of feeling, acting, and being were different and better. (Uhrmacher, 2010). They had been positively transformed by this aesthetic approach to teaching science.

TIME FOR EXPERIENCING, REFLECTION, AND CRITICISM

No resource is more precious to a teacher than time. Of the many lessons Elliot Eisner believed educators could learn from the arts was "the importance of paying close attention to what is at hand, of slowing down perception so that efficiency is put on the back burner and the quest for experience is made dominant" (2002, p.

207). Creating opportunities for students to "notice what there is to be noticed" (a key feature of Maxine Greene's conceptualization of aesthetic education) is a time investment well worth making. Crystal discovered that her autistic students in particular *excelled* at noticing. For the first *time* in their young lives, science was made accessible and available to these very special students. For example, they reveled in being allowed to draw in exact detail the particulars of the scales printed on measurement equipment. They loved the practice of measuring things. Sky challenged her students first to be still and then to see—to notice interactions on their school's nature trail between biotic and abiotic factors and then to record those observations through writing and drawing. Through performance and humor, Marsh appealed to his students to collaboratively help a hungry crustacean design a simple machine to retrieve a Bo Jangles chicken leg from the bottom of a trashcan. He allowed time for students to create and share videos which told the stories of their imagined solutions to the poor crustacean's problem.

Throughout this participatory action research, students created science *models* which were also *art* projects and products. Recall that throughout this book, art has been conceptualized as it was *before* it was subsumed by the wave of empiricism that emerged in the 18th century: "as any act of making that involves intellectual judgment" (Taylor, 1964, p. 46). The time investment for making/creating *science*-centered *art* projects also paid off in higher test scores and reported feelings of affirmation by teachers and students. In the end, what was *significant* was that now science *mattered* to previously struggling students.

In order to maximize the benefits associated with teaching science as aesthetic inquiry, time must also be allocated for student reflection and feedback. At regular junctures, students should be asked to weigh in on what they are feeling about how the learning is going (or not). If they say they're not learning, then they should be invited to suggest ways that might work better for them. At every opportunity, students should gain practice in the art of constructively criticizing each other's work so that each is challenged to perform at his or her highest level possible. In all these ways, an aesthetic approach encourages and supports a democratic, caring, and inclusive learning environment.

DRAWING ARTIFACTS AND ASSESSMENT

The arts artifacts generated by teacher participants and their students were analyzed inductively, using both objective and subjective criteria. The degree to which the participating teacher or student correctly made visible his or her understanding of the standards-based content was the *objective* standard of analysis. For example, during our group meetings and on our Project Draw for Science wiki discussion board, we spent a significant amount of time debating different ways to draw a given science idea. We most definitely adopted the mindset and practices of engineers in so doing. We collaboratively generated ideas, sketched them out, designed and tested each lesson, after which we reflected on what worked and what had not. We agreed that because so many science teachers have students who are reading below grade level or speak little English, at the very least, the creation of images should become a normative component of teaching/learning of all new science vocabulary. Through her innovative use of technology and her application of the pictorial superiority effect, Maya found a way for the profoundly dyslexic Muhammad to *show* what he knew when telling or writing the same was not possible.

The more *subjective* appeal of the drawing artifacts was important on multiple levels. For example, teachers learned that more artistically gifted students can "prettify" their drawings (explanatory models) to mask the fact that his or her understanding is incorrect or incomplete. Consider the following explanatory drawing about photosynthesis, which is aesthetically pleasing but conveys the misconception that *flowers* are producing food for the plant:

The take home lesson here was that teachers who want to use drawings as performance assessments will need to have constructed criterion-based rubrics. If the drawing performance assessments are summative in nature, both students and parents should be provided with the evaluation rubric ahead of time. The rubric should also clearly communicate that no one drawing style will be privileged over another. (No drawing talent is required.) A choice of drawing performances should be offered. Finally, in the rare event that a student demonstrates significant drawing anxiety (or fine motor difficulties), he or she should be given an alternative way to re-present their learning and evolving science literacy.

AN EVIDENCE-BASED MODEL

Adopting Eisner's (1976) educational connoisseurship and criticism approach in my evaluation practice and data analysis, I set out to determine what specific teacher behaviors (and concomitant teacher/student learning outcomes) might be characteristic of Science Teachers Who Draw. In doing so, I was also honoring Tochon's (2013) appeal to create a "grammar of teacher actions" which signified aesthetic teaching. My search for emerging patterns in the data was guided by seven research questions, each examining a different feature of this participatory inquiry: 1) **epiphanies**, or new ideas and understandings for all stakeholders; 2) **innovations**, which teacher participants might devise and employ as they explored the use of drawing to plan, design, produce, and ultimately enact teaching performances; 3) **connections,** which teacher researchers might make with self, content, researcher, student, parents, and administrators as a result of this metasemiotic curriculum inquiry; (4) potential conflicts or **obstacles** that might prevent or get in the way of a teachers' achieving their objective of teaching science through drawing; (5) the levels of **support** afforded to teachers as they engaged in this new pedagogy; (6) the form and content of drawing **artifacts** generated by the researcher, participating teachers and students as, together, we/they explored the teaching and learning of science through drawing; and 7) the **quantitative evidence**, if any, of improved student achievement or level of engagement as teachers implemented the practice of teaching science through drawing. For this book (and my dissertation) I selected only those teaching/learning performances which were clearly liminal for each teacher *and* which addressed the research questions. Then, I set out to create through performance narratives these teachers' *mystories* (Turner, 1982), which were situated in their classrooms with their own students. I intentionally did *not* seek to generalize the data or present or imply that there is one best or right way to teach science through drawing. Rather, I chose to invite my readers to experience the performance of the research so that they might draw their own conclusions.

Instead of re-presenting the thoughts and feelings of the narrators in paraphrased, linear summaries, I chose to *poetically* transform the most evocative of my interviews with my participants. I discovered the technique of poetic transcription in my reading of Lunsford Mears' (2009, 2010) *gateway approach*: the presentation of data in ways that create what Corrine Glesne (1997), has called the "third voice that is neither the interviewee's nor the researcher's but a combination of both" (p. 183). I liken the poetic distillation of the data to the process of gesture drawing, whose purpose it is to get down in as few lines as possible a drawing that communicates the *essence* of a thing. In the final re-presentation of the data, I adopted a *mixed-media* approach, creating a collage of ethnopoetics, Readers' Theatre scripts, and visual artifacts meant to evoke the distinct and unified voices of the performers. My intention was that the reader/audience would be able to enter the data and experience two performances in one—that which was vicarious and that which is fully present as the performative text was either read or enacted. Through my art, I hoped to aesthetically bridge my objective data analysis to the more human story of what happened in our research. I hoped that you might feel connected to our struggles, our epiphanies, our accomplishments, and our failures. I hoped to create the sense of *community* with my readers I had experienced with these teacher researchers and their students.

How does a Science Teacher Who Draws look or behave? This is a very difficult question to answer. No one teacher with whom I have worked has ever employed in the same way the arts skills I have helped them to develop. This is the nature of creative, emergent, arts-based methodologies and aesthetic inquiry. Any behavioral model or proposed "grammar of teacher actions" (Tochon, 2013) I might propose could not possibly be all inclusive. All models, like art, are open to alternative interpretations. And *yet*, based on the case studies from this research, my literature review, and twenty five years of field testing these methods, I believe I can confidently state that a Science Teacher Who Draws would employ as many of the following *aesthetic* teaching practices as possible:

1. Celebrate each student's intrinsic worth, value, and intuitive knowledge in a CARING learning community.

2. Slow down to allow TIME for students to perceive and to NOTICE what there is to be noticed and gives students time to draw and re-draw their EMERGING understanding of science phenomena and practices.

3. Develop DEEP enough SUBJECT MATTER CONTENT KNOWLEDGE to detect your own as well as student misconceptions.

4. Create VISUAL EXPLANATORY DRAWINGS of Big Ideas to formatively assess your own pedagogical content knowledge *prior* to teaching.

5. Guide instruction toward the generation of drawings which VISUALLY TELL THE STORY of emerging science understanding.

6. Do not privilege any single mode of drawing; rather CELEBRATE all INDIVIDUAL drawing styles— from the naïve and iconic to the realistically representational.

7. Create many OPPORTUNITIES for students to CRITIQUE each other's drawings in ways that are POSITIVE and SUPPORTIVE of one another.

8. View teaching and learning as CO-PERFORMANCES, which build CONNECTIONS to SELF, OTHERS, AND CONTENT.

9. Be FLEXIBLE enough to change your teaching 'script' when unexpected (surprise) learning opportunities arise and make sure to INCLUDE STUDENTS in the "editing"

process.

10. COLLABORATE and share your innovative practices with other colleagues so that new learning can be grafted onto an ever-evolving model.

I invite you to reflect upon how you would prioritize these teaching behaviors in terms of their potential impact on student learning. Every teaching/learning situation is different and also the same, depending on your perspective.

Conclusion

> *Oh chestnut-tree, great-rooted blossomer,*
> *Are you the leaf, the blossom or the bole?*
> *O body swayed to music, O brightening glance,*
> *How can we know the dancer from the dance?*

— from *Among School Children,* by William Butler Yeats

Like all science educators, I have struggled with how best to serve in the quest to achieve improved science literacy for American citizens at a time in our history when we have become a nation of poor or reluctant readers. Always, the task comes down to making science matter. I have proposed that teaching science as aesthetic inquiry can enable teachers to build connections between themselves, their students, and science content. Aesthetic inquiry requires a caring, empathic, *purposeful* engagement with the arts *and* the feelings to explore content in emergent ways that may lead to new and unanticipated learning destinations. As *choreographers*, the teacher creates performances wherein the dancer is indistinguishable from the dance, just as the trunk and leaves of the tree are part of the same organic whole.

The science teacher who adopts an aesthetic paradigm becomes skilled at matching art form with specific science content. In my master's thesis, I developed a learning progression through which the capacity for teaching science through the arts might be developed. I proposed that teacher learning would progress through four sequential phases: 1) an orientation to the nature of the creative process shared by both scientists and artists; 2) specific training and instruction in the arts area to be integrated into the teaching of the science content; 3) creative exploration of science themes through specific artistic means of expression; and 4) sharing and evaluation of arts products and discussions of the understandings arrived at through their creation. I employed this same paradigm for the participatory action research upon which this book has been based. The art practice was drawing; the science, those state standards which teacher participants were required to teach.

I began this book with the question *What If?* I now ask *What If* S.T.E.M. and Arts educators worked together to *cross-train* each other in ways that aligned with *both* their national standards? Early gatherings might focus on the many ways in which the new visual arts standards (available online through the National Art Education Association), overlap in both theory and in practice with the *Framework for K-12 Science Education*. There are many interconnections between the domains. For example, at the core of the art standards are four *Artistic Processes:* Creating, Presenting, Responding, and Connecting. Within each of these four artistic processes are *Enduring Understandings*, or assumptions underlying those practices. On the arts function of *connecting*, Stewart (2014) wrote that art making is "an investigative process, recognizing and using inquiry methods of observation, research, and experimentation as a means for exploring their own evolving interests and concerns as well as for constructing new knowledge and insights" (p. 10). No authentic *science* model can be created without employing very similar practices. Both the practice of science and art require critical thinking, problem solving, communication and collaboration, four skills deemed by the Partnership for 21st Century Skills, (www.P21.org) as being essential for success in work and in life.

Jacob Bronowski (1965) has called on us to "use all our faculties to the full—to assimilate with the scientist's brain, the poet's heart, the painter's eyes" (p. 20). In the Taoist philosophy, yin and yang produce all that comes to be. Science and art, yin and yang, may be seen as complementary expressions of insight, intuition, and discovery. When science teachers make time and space for the creative expression and re-presentation of substantive content knowledge through arts practice, deep, meaningful learning experiences can be achieved. In addition, many engineering and design practices are employed. Bring the language of mathematics in there, and you have S.T.E.A.M., or S.T.E.M. and the Arts. What matters is *not* what we call these teaching/learning experiences, but that they happen!

My teacher research participants and I determined that drawing is a powerfully communicative medium for those students who are poor or struggling readers. By teaching science as aesthetic inquiry through drawing, they successfully disrupted toxic patterns of failing. Every single breakthrough was cause for celebration. Previously failing, marginalized students stated that they now actually *cared* about science class. Moreover, they reported that drawing in science helped them better remember the content for tests. Their grades improved. Buoyed by new success, previously disengaged students eagerly asked to draw so that they could learn even more science. To our amazement, they even began creating *extra* science related drawings at home.

Each teacher in this study adopted what Baldiali and Hammond (2002) have called an "inquiry stance", which is essentially a "scientific" strategy of identifying a problem, designing a solution, testing it, making judgments about its effectiveness, and then modifying practices until some improvement is achieved (in Green and Johnson, 2010, p. 19). Their results were so impressive that three teachers were asked to present their findings at district level professional development workshops. Teacher *leaders* were born.

> The forests speak out, the oceans beckon, the sky calls us forth, the plants want to share their story, the mind of the universe is open to all of us, the planet wants to instruct. Educators, through

their methods, can open wide the doors to this wonder (LePage, 1987, p. 180).

To our research team, improved test scores were not what truly mattered (though, of course, we were thrilled with this outcome). Marsh, Sky, Maya, Bones, and Eartha (working underground) discovered that nearly all of their previously struggling students (mostly all poor readers) were *thriving*. Students reported feelings of frank joy, freedom, and empowerment, a sign that aesthetic energy was being transferred throughout the learning environment. Students with special needs, especially those who were dyslexic and/or identified with autism spectrum disorder seemed particularly to benefit. I hope to continue my research on ways in which not only drawing but also other aesthetic media may be employed to empower special needs students to fully, equitably, and *joyfully* participate in the learning of science.

Throughout the research study, my team of teacher researchers continually surprised me with their highly innovative lessons, as they and their students generated hundreds and hundreds of drawing artifacts which bore witness to the multifaceted meaning-making experienced by all who took part in these performances. In the final reflection phase, participant teachers bore witness to what "being there" was like for them as they experienced the dual roles of change agent and subject of change. Such reflection required each person to respond from deep within themselves, especially since what they were trying to create was a work of art that took the form of both teaching and learning. Greene (1978) has called for "wide-awakeness" in education (p. 4). One student in the study voiced the same sentiment: *Science makes me feel great, because it's the only class I get to draw in beside art that has imagination. I feel like I want to come to science every day. It makes me feel alive!*

Glossary

Art: Throughout this book, art has been conceptualized as it was *before* it was subsumed by the wave of empiricism that emerged in the 18th century—as any act of making that involves intellectual judgment. Art is the medium through which human beings develop their intuitive and creative potential, regardless of the domain inside of which (like science or teaching) it is practiced.

Art / Science project: A creative design task undertaken by a student or group of students which uses arts practice to apply or explain science concepts or phenomena.

Aesthetics: In this book, I have employed Baumgartner's original *sense*-based conceptualization of this term together with Dewey, Greene, Eisner's and Siegesmund's more ontological characterizations of aesthetics as being purposefully empathic and caring, interactive, connective, feelful, and centered around the practice of keen noticing the particular qualities of something, whatever that may be.

Aesthetic energy: A renewable form of human energy characterized by the presence of openness, care, empathy, emotional connection, sustained acts of noticing and awareness, and flexible, creative ontologies and epistemologies.

Convection: A form of heat energy transfer within fluids that takes place *within* the movement of the heated substance itself. Convection is triggered by a heat source, which heats the molecules of a fluid medium until they become less dense and begin to rise. As the heat energy is given off to the surroundings, the fluid medium becomes more dense and sinks back down toward the heat source, only to begin the process again. Such convection "cells" can be maintained as long as there is a heat source. In this book, the teacher has been analogized as the heat source and the learning process as a convective cell.

Contour drawing: A slowly rendered drawing made by allowing the pencil or pen to move on the drawing surface at the same rate that the eyes follow the inner and outer edges (contours) of the object under study. Such a drawing is often rendered without lifting the drawing tool at all. A "blind" contour drawing is completed by turning away from the drawing surface (or covering the hand with another sheet of paper) so that the brain cannot pass "judgment" on the drawing while it is being created. Such drawings are excellent ways to practice intense "noticing" and to discover the relationship between part to whole.

Educational criticism and connoisseurship: These two practices were conceptualized by Elliot Eisner (1976) as a new way of conducting educational evaluations (as opposed to the five minute checklist for specific objective criteria). The process was to be carried out by one functioning as a *connoisseur,* who first engaged in privately appreciating and noting the particular qualities of a teaching/learning performance. The more public act of criticism required that they experience be written about in a way that clearly communicated the essence of those experiences. According to Eisner, a valid criticism could only be written by one who was a connoisseur. For example, because I have never taught German and have only a rudimentary command of the language, I could not write a valid criticism of a German teacher's lesson, even though I might enjoy the performance.

Edusemiosis: As I understand this term, the entire enterprise of education either functions (or does not) depending on whether the sign meaning-making process (semiosis) communicates information in such a way that it is mutually understood by teacher and student. Conceptualized in this manner, students may be "starved" if the meaning of signs and symbols is not made accessible to them. As a result, they will become marginalized within the ecology of the classroom and forced to develop their own entrenched niches.

Energy transfer: This term is used in both a physical and metaphysical way in this book. In physics, energy can be put into an object, causing it to be *moved* in some way. This is the scientific conceptualization of "work". Metaphysically speaking, students can be inspired or "moved" into a higher state of attention and engagement and thus will likely produce more "work". Teaching science as aesthetic inquiry is pedagogy in which both the physical and metaphysical aspects of the educative process are deemed important educational research variables.

5E Learning Cycle: A sequence of steps for classroom science inquiry developed by Roger Bybee and a team from the the Biological Science Curriculum Study (BSCS), though which a student might construct their own understanding of new ideas: 1) **Engage** (Interest is captured and prior knowledge established.); **Explore** (Students are guided through inquiry activities with teacher questions designed to help students to make their own inferences and conclusions; 3) **Explain** (Teacher formatively assesses student understanding by asking them to re-present their understanding in some way. This is followed by reflection to disrupt misconceptions.); 4) **Extend** (Students apply their knowledge to a different but similar situation through reasoned argument.); **Evaluate** (Teacher assesses student understanding either formatively, throughout the cycle, or summatively, at the conclusion of a lesson.)

Gesture drawing: This type of drawing involves a practice which is opposite that of the painstaking methods of contour drawing. Instead, one draws (to quote Nicolaides) "incessantly and furiously". With the gesture drawing, one tries to capture the essence, spirit, or movement of the subject with bouts of intense, brief sketching. Placing extreme time restrictions (say less than 30 seconds) on the practice will force one to see more carefully.

Joint productive activity: Considered best practice in pedagogy, this teaching/learning experience is co-performed by teacher and students who are united in creating a common learning product or achieving a stated learning goal.

"Know"tation: A play on the word *notation*, I created this term for two reasons: 1) to communicate the reflexive, generative nature of science as a domain of knowledge and practice. 2) to encourage both teachers and students to visually organize their understandings of science concepts, interactive experiences, or phenomena according to the principles of graphic design. The goal of the "Know"tation is to show on a single page, through an aesthetically pleasing combination of words (typography), drawings, and white space, the story of a teaching/learning experience in science so that an uninformed viewer can "read" that story. In every case, "Know"tations should be collaboratively assessed for the presence of misconceptions and revised until the visual "story" as closely as possible re-presents the most accurate level of "knowing" about that science subject. "Know"tations can be created at every stage of the 5E learning cycle.

Metasemiosis: As explained by Tochon (2013), this is the singularly human phenomenon of thinking about how it is that we make meaning. If we are educators, unless we reflexively and recursively examine our ways of effecting semiosis in our classrooms, we may inadvertently leave students out of the learning equation. If we employ a multimodal, multi-media approach to teaching, we are likely to make learning accessible to a more diverse group of students.

Mirror neurons: These are specialized brain cells that fire whether we see someone else perform an action or do the action ourselves. (See Winerman, 2005 article in References.) If students are making new neuronal connections which enable them to recall the actions and behaviors of their teachers, might the mirror neuron phenomenon be the reason teachers teach the way in which they have been taught? Since the brain has been deemed to be "plastic", how can modeling aesthetic teaching practices in science communicate the idea that science and art are complementary rather than distinct, or even mutually exclusive domains of education?

Mutualism: Typically discussed in the ecological sciences, mutualism describes a type of relationship between organisms of two entirely different species (and often living requirements) in which both species benefit in some way. For example, fragile photosynthetic algae (called zooxanthellae) live within the rigid calcium carbonate skeletons of reef-building corals. The algae provide food and oxygen (and gorgeous color), while the coral provide shelter and defense. In this book, I have proposed that there exists between science and art a natural *mutualistic* relationship, each with the potential to nourish and enhance the other within the ecology of the classroom and laboratory.

Narrative performative drawing: This is a dramatically scripted, staged, contextualized, and co-performed, live-action drawing which, through simple line drawings and only key words, *shows*, rather than tells the story of the science concept(s) being featured as the "main character(s)". Marsh's performance of "Chilli the Mutant Crab and the Bo Jangles Chicken Leg" was an example of a narrative performative drawing.

Neuroplasticity: This recent finding by neuroscientists that the brain responds to outside influences by changing both its cell structure and function (via altered protein synthesis and the laying down of synaptic connections between neurons) is both good and bad news for educators. Doidge (2007) has explained that once a particular "plastic" change occurs in the brain and becomes well established, it can make it difficult (but not impossible) for other changes to occur. According to Doidge, the phenomenon has the power to produce both flexible and rigid behaviors. I posit that a multimodal, emergent, aesthetic approach to teaching will increase the likelihood that the students' learning in that environment will display more flexible, less linear approaches to problem solving and meaning-making.

Protein synthesis: The making of proteins is the very stuff of life, living, learning, and therefore edusemiosis. They function in transport, building, repair, and regulation. The "blueprints" for making proteins are coded as "genes" on long strands of molecules called DNA, which are bunched up on chromosomes inside the nuclei of each cell. Erik Kandel won the Nobel prize in medicine for determining that when we learn, we alter which genes are "turned on" in our brain cells (neurons), and therefore, which proteins are made. As long ago as 1888, Freud proposed that "neurons that fire together wire together" (Doidge, 2007, p. 223).

Semiosis: The meaning-making process, contextually situated in lived experience. In Peirce's theory of signs, semiosis depends on the triadic interaction between an object in the world, a sign or symbol that re-presents that object (or phenomenon), and the way(s) a person in a given culture or situation interprets that contextual sign.

Semiotics: The science and philosophy of the phenomenon of meaning-making, or semiosis.

Science model: As conceptualized in this book, a science model is a theoretical re-presentation of a concept or phenomenon based either on physical evidence or reasoned conjecture. The relative "value" of a scientific model is determined by its ability to predict measurable outcomes, explain past events, or inform future research. Many of the most important science models came about in moments of epiphany (the so-called *Eureka* moments) after periods of sustained aesthetic noticing of the particular qualities of the concept under study.

S.T.E.M: An acronym for Science, Technology, Engineering, and Mathematics as areas of academic study and professional practice.

S.T.E.A.M: An acronym for a curriculum arising from the research of Georgette Yakman which proposes that science and technology be "interpreted through engineering and the arts, all based in the language of mathematics" (http://www.steamedu.com/).

TSTA: An acronym for Teaching Science through the Arts, a model I developed which employs arts practices to aesthetically connect students to science content in meaningful and relevant ways.

S.T.E.A.M School: A hypothetical action research setting in which teachers with fluency in either S.T.E.M. or the Arts (or both) would form a curriculum building team. First, they would "cross-train" each other and then together create innovative and integrative curriculum for teaching and learning that is congruent with both S.T.E.M. and the Arts standards. Such "schools" could form and function within all levels of K-20 education, including and especially in teacher education programs. An underlying assumption would be that all curricula would be culturally relevant.

S.T.E.A.M. Camp: A hypothetical action research setting in which teachers who want to learn how to implement S.T.E.A.M. curricula interact with a "certified" S.T.E.A.M. Team (see above) and groups of students to assess the impact of teaching/learning in this integrated manner.

S.T.EA.M. Expo: A hypothetical, celebratory event featuring the public display of artistic interpretations of S.T.E.M. concepts and phenomena in such a way that an aesthetic connection is created and the audience *feels* the wonder the artist experienced in creating the work.

Symbology: In this book, I have used this term to represent the complex system of symbols (visual, verbal, and embodied/ canonical and expressive) through which students associate learning science content with feelings ranging from frank boredom to felt freedom and excitement. Eisner (1976) has stated that "symbols possess in their form the expressive content to which they are related" (p. 137). Because science symbology is so often not taught in conjunction with the stories that created them, their meaning can be easily lost in translation. An aesthetic approach to teaching science symbology foregrounds both story and expression and fosters deeply associative, collaborative learning.

Epilogue

Notes for the Qualitative Researcher: ABER

In the conclusion of *The Enlightened Eye: Qualitative Inquiry and the Enhancement of Educational Practice*, Eisner (1991) describes Arts-Based Educational Research (ABER) as research about "teaching practices... that use resources from the visual world, from music and dance, and from poetry and literature [to] enable children to grasp what cannot be revealed in text" (p. 246). When an arts-based approach to education is employed, as Eisner explained in *The Arts as Creation of Mind*, "the ends are held flexibly...and need not be and often cannot be specified with any degree of certainty in advance" (p.206). Eisner devoted his entire career to endeavoring to describe ways in which artistic ways of teaching might be more fully accepted, explored and enacted. He actively called for university professors and researchers to go into classrooms to discern those "qualities" which characterized artistic forms of teaching and multiple ways of knowing and being human beings and then to re-present their data in ways that authentically and evocatively capture the complexity of classroom interactions. He vehemently called for schools, which opened both space and time for teachers to evaluate each other's performances through shared productive criticism and decried the imposed and predominant isolation of teachers one from the other.

As director, producer, and supporting actress in this improvisational performance, I maintained several journals and sketchbooks of my own reflections about this experience and scribbled analytic memos in the margins of my interview transcriptions. Quite often, my journal reflections took either poetic or graphic form. With each pass through the data, I was actively searching for each teacher's *mystory*, described Turner (1982) as progressing through a series of "liminal" experiences, including "epiphanic moments", "turning point experiences", and times of personal turmoil and conflict. (p. 34-35). I intentionally tried to elicit teachers' stories about their "epiphanic moments" (later coded as EPIPHANIES) during our interviews and group meetings with the idea that such stories could become what Saldaña (1999) described as "the juicy stuff" (p. 61). Occasionally, during the study, a teacher's lesson would completely backfire, experiences we also classified as EPIPHANIES because there was much to learn from them. One theory, which emerged early on, was that an epiphany of some sort had to occur before any kind of transformation could follow.

In keeping with the epicentric theme of transformation, participants' reflections, responses, and teaching performances were evaluated in terms of how they described or enacted the process and practice of meaning-making (semiosis) in the science classroom as together we explored the implications of the research questions. The constant comparative method (Glaser & Strauss, 1967) was used for the construction of emergent themes about seven areas of inquiry, which were coded as follows: EPIPHANY, INNOVATION, CONNECTION, OBSTACLES, SUPPORT, ARTIFACTS, and QUANTITATIVE DATA. The first coding cycle involved *in vivo* protocol coding, with categories derived from the seven research questions. For example, depending on whether the *r*esearcher, *p*articipating teacher or *st*udent was having an EPIPHANY, this recurring event was coded as E*r*, E*p*, or E*st*. Likewise, all of us—myself (researcher), the participating teachers, the students—devised our own drawing-to-teach/learn INNOVATIONS (hence the I*r*, I*p*, and I*st* codes). The OBSTACLE code was treated descriptively by individual occurrence in analytic memos and included events like resistance to or fear of drawing on the part of the teacher or student, an overcrowded classroom, time constraints, etc. The CONNECTION code was used to locate in the data examples of connections with *content, self, researcher, student, administrators* and *parents*. These were coded as Cc, Cs, Cr, Cst, Ca, and Cp respectively. During the first and second cycles of coding, I generated analytic memos in the margins of the coded transcriptions. A third coding cycle involved the poetic distillation of *in vivo* codes, with the goal of symbolizing the interactions between all who took part in this story as we collaboratively explored the research questions.

During my study, I put into place many opportunities for validity testing. Becker and Geer (1957) claimed that long-term participant observation is the best method for providing complete qualitative data. By the end the study, I had been working with these teachers for nearly 10 months, employing multiple methods of data collection, which could be triangulated against one another. In this ten months, through intensive one-on-one interviews, emails, and many hours of group collaboration and problem solving, I collected a body of very rich data, about which I generated thick description, through which I attempted to incorporate "the intentions of the actors and the codes of signification that that give their actions meaning for them" (Maxwell, 2013, p. 138). The most important validity test, however, was that of seeking each participant researcher's assessment of the degree to which I had credibly captured and portrayed the meaning and significance of their perspectives on a set of events and experiences (Lunsford Mears, P. 25). Each participant was given copies of transcribed interviews as well as my written interpretations of all shared experiences and artifacts.

From a triangulation of my multiple data sources, and as the analysis evolved, I constructed a three-dimensional portrayal of participants by distilling the transcripts of interviews in such a way as to "capture the essence of the experience" (Lunsford Mears, 2009), all the while carefully noting what each participant revealed about his or her perceptions or constructed meanings from interviews, observations, and the artifacts they and their students produce. Running through the final performative text was my own interpretive and dialogic voice as a kind of autoethnographer-director, a dramatic device whose purpose was to unequivocally establish my own reflexivity, described by Pillow (2003) as the ways in which "who I am, who I have been, who I think I am, and how I feel affect data collection and analysis" (p. 176). Ultimately, my reflexive analysis of the data corpus was assimilated and described in the poetic narrative of the performance ethnography and illustrated with arts artifacts, which were used to support the unfolding performance.

References

Adjibolosoo, S. (1995). *The human factor in developing Africa.* Westport, CT: Praeger.

Ainsworth, S.E (2006). DeFT: A conceptual framework for learning with multiple representations. *Learning and Instruction, 16*(3), 183-198.

Ainsworth, S. (2008). The educational value of multiple representations when learning complex scientific concepts. In J. K. Gilbert, M. Reiner, & M. Nakhlel (Eds.), *Visualization: Theory and Practice in Science Education* (pp. 191-208). New York: Springer.

Ainsworth, S. (2011). Drawing to learn in science. *Science , 333*, 1096-1097.

American Association for the Advancement of Science (1989). *Science for All Americans: Project 2061.* Retrieved from http://www.project2061.org/publications/sfaa/online/sfaatoc.htm.

Anderson, L. (1931, 1970). *Pestalozzi.* New York, NY: AMS Press.

Anderson, L., & Krathwohl, D. R. (2001). *A taxonomy for learning, teaching, and assessing: A revision of Bloom's taxonmy of educational objectives.* New York: Longman.

Aubusson, P. J., Fogwill, S., Barr, R. & Perkovic, L. (1997). What happens when studentsdo simulation role play in science? *Research in Science Education, 27*(4), 565-579.

Author, K. R. (2009). Science through drama: a multiple case exploration of the characteristics of drama activities used in secondary science lessons. *International Journal of Science Education, 31*(16), 2247-2270.

Baldiali, B., & D. Hammond. (2002). The power and necessity for using inquiry in a PDS. Paper presented at the Professional Development Schools National Conference, Orlando, FL.

Barba, R. (1995). *Science in the multicultural classroom: A guide to teaching and learning.* Boston: Allyn and Bacon.

Barab, S. A., & Duffy, T. M. (2000). From practice fields to communities of practice. In D. H. Jonassen & S. M. Land (Eds.), *Theoretical foundations for learning environments* (pp. 35-56). Mahwah, NJ: Lawrence Erlbaum.

Berger, J. (2011). *Bento's sketchbook: How does the impulse to draw something begin?* New York, NY: Pantheon Books.

Black, P., & William, D. (1998). Assessment and classroom learning. *Assessment in Education, 5* (1), 7-74.

Borko, H., & Putnam, R. T. (1996). Expanding a teacher's knowledge base: A cognitive psychological perspective on professional development. In T. R. Guskey & M. Huberman (Eds.) *Professional development in education: New paradigms & practices* (pp. 35-66). New York: Teachers College Press.

Bourdieu, P. (2003). *The forms of capital*. In A. H. Halsey, H. Lauder, P. Brown, & A. S. Wells (Eds.), *Education: Culture, economy, society* (pp. 46-58). Oxford, UK: Oxford University Press.

Boutte, G. (1999). *Multicultural education: Raising consciousness*. Belmont, CA: Wadsworth.

Boutte, G. S., & Hill, E. L. (2006). African-American communities: Implications for culturally relevant pedagogy. *The New Educator, 2*, 311-329.

Bresler, L. (2006). Toward connectness: Aesthetically based research. *Studies in art education, 48*(1), 52-69.

Bronowski, J. (1965). The discovery of form. In G. Kepes (Ed.), *Structure in art and science*. New York: Braziller.

Brown, J. S., Collins, A., & Duguid, P. (1989). Situated cognition and the culture of learning. *Educational researcher, 18*(1), 32-42.

Bybee, R.W. et al. (1989). *Science and technology education for the elementary years: Frameworks for curriculum and instruction*. Washington, D.C.: The National Center for Improving Instruction.

Carnine, L. & Carnine, D. (2004). The interaction of reading skills and science content knowledge when teaching struggling secondary students. *Reading and writing quarterly. 20*, 203-218.

Catterall, J. (1998). Does experience in the arts boost academic achievement? A response to Eisner. *Art Education, 51*(4), 6-11.

Charlesworth, M. J. (1957). *Aristotle on art and nature.* Auckland University College Bulletin No. 50, Philosophy Series No. 2. Auckland: University College.

Chien, J. P. (2006). From Animals to Humans: Uexküll's Umwelt as Read by Lacan and Canguilhem. Retrieved from http://www.concentric-literature.url.tw/issues/Who%20Speaks%20for%20the%20Human%20Today/3.pdf

Coleman, J. S. (2003). The forms of capital. In A. H. Halsey, H. Lauder, P. Brown, & A. S. Wells (Eds.), *Education: Culture, economy, society* (pp. 80-95). Oxford, UK: Oxford University Press.

Conquergood, D. (1998). Beyond the text: Toward a performative cultural politics. In S. Dailey (Ed.) The future of performance studies: Visions and revisions (pp. 25-36). Annandale, VA: National Communication Association.

Cromley, J. (2009). Reading achievement and science proficiency: International comparisons from the programme on international student assessment. *Reading psychology. 30*(2), 89-118.

Darwin, C. (1859). *Origin of the Species.* The United Kingdom: John Murray.

Deely, J. (2001). *Four ages of understanding: The first postmodern survey of philosophy from ancient times to the turn of the twenty-first century.* Toronto: University of Toronto Press. Retrieved from http://tinyurl.com/bw84unh

Delpit, L. (1988). The silenced dialogue: Power and pedagogy in educating other people's children. *Harvard Educational Review, 58*, 280-298.

Denzin, N. K. (1997). Performance texts. In W. Tierney & Y. Lincoln (Eds.), *Representation and the text: Re-framing the narrative voice.* Albany, NY: SUNY Press.

Denzin, N. K. (2003). *Performance ethonography: Critical pedagogy and the politics of culture.* Thousand Oaks, CA: Sage.

Denzin, N. K. (2009) 'A critical performance pedagogy that matters'. *Ethnography and Education, 4*(3), 255-270.

Dewey, J. (1934). *Art as experience.* New York, NY: Capricorn Books.

Dewey, J. (1938). *Experience and education.* New York: Kappa Delta Pi.

Dinkins, C. S. (2005). Shared inquiry: Socratic-hermeneutic interpret-viewing. In P. Ironside (Ed.), *Beyond method: Conversations in healthcare research and scholarship* (pp.114-147). Madison, WI: University of Wisconsin Press.

diSessa, A. A. (2004). Metarepresentation: Native competence and targets for instruction. *Cognition and instruction, 22*(3), 293-331.

Doige, N. (2007). *The brain that changes itself.* New York: Penguin.

Dorion, K. (2011). A Learner''s Tactic: How secondary students'' anthropomorphic language may support learning of abstract science concepts. *Electronic Journal of Science Education, 15*(2).

Driver, R. (1989). Students' conceptions and the learning of science. *International Journal of Science Education,* 11, 481-490.

Dwyer, M. C. (2011). Reinvesting in Arts Education: Winning America's Future through Creative Schools. *President's Committee on the Arts and the Humanities.*

Ehrenfeld, T. (2011). Reflections on mirror neurons. *Observer,* 24(3), March. Retrieved from http://www.psychologicalscience.org/index.php/publications/observer/2011/march-11/reflections-on-mirror-neurons.html

Eisner, E. W. (1972). *Educating artistic vision.* New York: Macmillan.

Eisner, E. W. (1976). Educational criticism and connoisseurship: Their form and function in educational evaluation. *Journal of Aesthetic Education, 10*(3/4), 135-150.

Eisner, E. W. (1991). *The enlightened eye: Qualitative inquiry and the enhancement of educational practice.* New York: Macmillan.

Eisner, E. W. (2002). *The arts and the creation of mind.* New Haven, CT: Yale University Press.

Eisner, E. W. (2009). What education can learn from the arts. *Art education, 62*(1), 6-9.

Ellison, R. (1963). What these children are like. September, 1963 lecture. Retrieved July, 2013 from http://teachingamericanhistory.org/library/document/what-these-children-are-like/.

Esrock, E. J. (1994). *The reader's eye: Visual imaging as reader response.* Baltimore, MD: John's Hopkins University Press.

Finson, K., & Pederson, J. (2011). What are visual data and what utility do they have for science education? *Journal of visual literacy, 30*(1), 66-85.

Frankel, F. (2010). *Picturing to learn.* Retrieved from http://www.picturingtolearn.org/

Freeman, C., Seashore, K., with Werner, L. (2003). *Models of implementing arts for academic achievement: challenging contemporary classroom practice.* Minneapolis, MN: Center for Applied Research and Educational Improvement.

Freire, P. (2004). *Pedagogy of hope.* London: Bloomsbury Academic.

Gage, N., & Berliner, D. (1991). *Educational psychology* (5th ed.). Boston, MA: Houghton Mifflin.

Glaser, B. G. & Strauss, A. L. (1967). *The discovery of grounded theory: Strategies for qualitative research.* New York: Aldine Gruyter.

Glesne, C. (1997). That rare feeling: Re-presenting research through poetic transcription. *Qualitative inquiry, 3*(2), 202-222.

Godin, S. (2012). *The Icarus deception: How high will you fly?* New York: Penguin.

Grant, G., & Grant, A. (2012). *Who killed creativity and how do we get it back?* New York, NY: John Wiley.

Green, S., & R. Johnson. (2010). *Assessment is essential.* Boston, MA: McGraw Hill.

Greene, M. (1971). Curriculum and consciousness. *Teachers College Record. 73*(2), 253-270. .

Greene, M. (2001). *Variations on a blue guitar: The Lincoln Center Institute lectures on aesthetic education.* New York and London: Teachers College Press.

Greeno, J. G., & Hall, R. P. (1997). Practicing Representation. *Phi Delta Kappan, 78*(5), 361-67.

Guskey, T. (1986). Staff development and the process of teacher change. *Educational Researcher, 15*(5), 5-12.

Hapgood, S., & Palincsar, A. S. (2006–2007). Where literacy and science intersect. *Educational Leadership, 64*(4), 56–61.

Heid, K. (2005). Aesthetic development: a cognitive experience. *Art Education*, September, 48-53.

Hubber, P., Tytler, R., & Haslam, F. (2010). Teaching and learning about force with a representational focus: Pedagogy and teacher change. *Research in Science Education, 40*(1), 5-28.

Huberman, M. (1995). Networks that alter teaching: Conceptualizations, exchanges and experiments. *Teachers and Teaching: Theory and Practice, 7*(2), 193-211.

Jewitt, C. (2007). A multimodal perspective on textuality and contexts. *Pedagogy, Culture, and Society, 15*(3), 275-289.

Johnson, C. (1955). *Harold and the purple crayon.* New York: Harper Collins.

Kelehear, Z. (2008). *Instructional leadership as art: Connecting ISLLC and aesthetic inspiration.* Lanham, MD: Rowman & Littlefield Education.

Kemmis, S., & McTaggart, R. (2005). Participatory action research: Communicative action and the public sphere. In Denzin, N. & Y. Lincoln, (Eds.), *The Sage Handbook of Qualitative Research Third Edition* (pp. 559-603). Thousand Oaks, CA: Sage Publications.

Klee, P. (2012). On modern art. In J. Sallis (Ed.), *Paul Klee philosophical vision: From nature to art* (pp. 9-23). Chestnut Hill, MA: McMullen Museum of Art at Boston College.

Koester Southgate, M. (1989). *Science and art together again.* Honolulu, HI: University of Hawaii, Unpublished master's thesis.

Koester Southgate, M. (2007). *Agnes pflumm and the stonecreek science fair.* (3rd ed.). Charleston, SC: Read for Science.

Koester Southgate, M. (2007). *No place like periwinkle.* (2nd ed.). Charleston, SC: Read for Science.

Koester Southgate. (2008). *Pond scum and agnes pflumm.* (2nd ed.). Charleston, SC: Read for Science.

Koester Southgate, M. (2011). *Agnes pflumm and the secret of the seven.* Charleston, SC: Read for Science.

Kozma, R. (2003). Material and social affordances of multiple representations for science understanding. *Learning and Instruction*, 13(2), 205-226.

Kull, K., Emmeche, C. & Favareau, D. (2008). Biosemiotic questions. *Biosemiotics*, 1, 41-55.

Ladson-Billings, G. (1994). *The dreamkeepers/Successful teachers of African American children.* San Francisco: Jossey-Bass.

Landau, F. (n.d.) Roots. Retrieved from https://faculty.unlv.edu/landau/roots120.htm

Lather, P. (1993). *Fertile obsession: Validity after poststructuralism.* Sociological quarterly, 34 (4), 673-693.

Latta, M. & Baer, S. (2013). Aesthetic inquiry: About, within, and through repeated visits. In T. Constantion and B. White (Eds.), *Essays on aesthetic education for the 21st century* (pp. 93-107). Rotterdam: Sense.

Lederman, N. G. (1992). Students' and teachers' conceptions of the nature of science: A review of the research. *Journal of research in science teaching, 29*(4), 331-359.

Lederman, N. (2007). Nature of science: Past, present, and future. In S. Abell & N. Lederman (Eds.), *Handbook of research on science education*. Abingdon: Taylor and Francis.

Lee, O., Fradd, S. H., & Sutman, F. X. (1995). Science knowledge and cognitive strategy use among culturally and linguistically diverse students. *Journal of Research in Science Teaching, 32*(8), 797–816.

Lemke, J. (2004). The literacies of science. Retrieved from http://www.jaylemke.com/storage/Literacies-of-science-2004.pdf

LePage, A. (1987). *Transforming education: The new 3 r's.* Oakland, CA: Oakmore House.

London, P. (1989). *No more secondhand art: Awakening the artist within.* Boston, MA: Shambhala.

Lortie, D. (1975). *Schoolteacher: a sociological study.* Chicago: University of Chicago Press.

Los Altos Writers' Roundtable. (1966). *Borrowed water: A book of American haiku.* Rutland, VT: Tuttle.

Loucks-Horsley, S., & Matsumoto, C. (1999). Research on professional development for teachers of mathematics and science: The state of the scene. *School science and mathematics, 99*(5), 258-271.

Louv, R. (2005). *Last child in the woods: Saving our children from nature-deficit disorder.* Chapel Hill, NC : Algonquin Books of Chapel Hill.

Lunsford-Mears, C. (2009a). *Interviewing for education and social science research: The gateway approach.* New York: Palgrave Macmillan.

Lunsford-Mears, C. (2009b). Researching complex and challenging issues: An approach developed for a qualitative study of the Columbine tragedy. A presentation to the 2010 Conference of the American Educational Research Association. Retrieved from academia.edu.

MacCurdy, E. (1941). *The notebooks of Leonardo da Vinci.* Garden City, NY: Garden City.

McGregor, D., Anderson, D., Baskerville, D., & P. Gain (2014). How does drama support learning about the nature of science: Contrasting narratives from the UK and NZ. http://www.esera.org/media/esera2013/Debra_McGregor _16Feb2014.pdf

Moses, R.P., & Cobb, C.E. (2001). *Radical equations: Civil rights from Mississippi to the algebra project.* Boston, MA: Beacon.

National Academy of Science (2013). All standards, all students: making next generation standards accessible to all students. *Next Generation Science Standards, Appendix D. Diversity and Equity.* Washington, DC. Retrieved from http://www.nextgenscience.org/sites/ngss/files/Appendix %20D%20Diversity%20and%20Equity%20-%204.9.13.pdf

National Center for Research on Teacher Education. (1991). Final Report: National Center for Research on Teacher Education. Michigan State University: East Lansing: MI. National Partnership for Excellence and Accountability in Teaching (NPEAT). "Principles of Effective Professional Development," *Research Brief.* Alexandria , VA: Association for Supervision and Curriculum Development, 2003). Vol. 1, No. 15.

Nelson, D., Reed, V., & Walling, J. (1975). Pictorial superiority effect. *Journal of Experimental Psychology: Human Learning and Memory, 2*(5), 523-528.

Nicolaides, K. (1961). The natural way to draw. Boston, MA: Houghton Mifflin.

Noddings, N. (2005). The challenge to care in schools: An alternative approach to education (2nd ed.). New York: Teachers College Press.

Norris, S. P., & Phillips, L. M. (2003). How literacy in its fundamental sense is central to scientific literacy. *Science Education, 87*(2), 224–240.

Nussbaum, M. (1997). *Cultivating humanity*. Cambridge: MA: Harvard University Press.

Oakes, J. (2005). *Keeping track: How schools structure inequality*. New Haven, CT: Yale University.

Paivio, A., & Csapo, K. (1973). Pictorial superiority in free recall: imagery or dual coding? *Cognitive Psychology ,2*,176-206.

Peirce, C.S. (1877).The fixation of belief. *Popular Science Monthly, 12,* 1-15. Also in Peirce Edition Project, Writings of Charles S. Peirce, Vol. 3: 1872-1878 (pp.242-256). Retrieved on June 5, 2007 from http://www.peirce.org/writings/p107.html

Perry, T., Steele, C., & Hilliard, A. III. (2003). *Young, gifted, and Black: Promoting high achievement among African American students*. Boston: Beacon.

Pestalozzi, J. (1894/1973). *How Gertrude teaches her children*. New York, NY: Gordon.

Pillow, W. (2003). Confession, catharsis, or cure? Rethinking the uses of reflexivity as methodological power in qualitative research. *Qualitative studies in education, 16*(2), 175-196.

Phillips, L. & R. Siegesmund. (2013). Teaching what we value: Care as an outcome of aesthetic education. In T. Constantion and B. White (Eds.), *Essays on aesthetic education for the 21st century* (pp. 221-234). Rotterdam: Sense.

Polanyi, M. (1966). *The tacit dimension*. New York, NY: Doubleday.

Pratt, H., & Bybee, R. W. (2012). *The NSTA Reader's Guide to a Framework for K-12 Science Education*. Arlington, VA: NSTA.

Rabkin, N., & Redmond, R. (Eds.). (2004). *Putting the arts in the picture: Reframing education in the 21st century*. Chicago, IL: Center for Arts Policy at Columbia College Chicago.

Richardson, L. (1997). *Fields of play: Constructing an academic life*. New Jersey: Rutgers University Press.

Roam, D. (2008). *The back of the napkin*. New York: Penguin.

Rolling, J. H. (2013). *Swarm intelligence: What nature teaches us about shaping creative leadership.* New York: Palgrave MacMillan.

Rolling, J. H. (2014). Artistic method in research as a flexible architecture for theory building. *International Review of Qualitative Research, 7*(2), 161-168.

Saldaña, J. (1999). Playwriting with data: Ethnographic performance texts. *Youth Theatre Journal, 13,* 60-71.

Scott, P. H., Asoko, H. M., & Driver, R. H. (1992). Teaching for conceptual change: A review of strategies. In R. Duit, F. Goldberg, & H. Niedderer (Eds*.), Research in physics learning: Theoretical issues and empirical studies* (pp.310-329). Kiel, Germany: University of Kiel.

Shulman, L. S. (1987). Knowledge and teaching: Foundations of the new reform. *Harvard educational review, 57*(1), 1-23.

Seashore, K. Anderson, A. & Riedel, E. (2003). *Implementing arts for academic achievement: The impact of mental models, professional community and interdisciplinary teaming.* Unpublished manuscript, Center for Applied Research and Education Improvement, College of Education and Human Development, University of Minnesota.

Shulman, L. (1986). Those who understand: Knowledge growth in teaching. *Educational researcher, 15*(4), 2-16.

Siegesmund, R. (2010). Aesthetics as a curriculum of care and responsible choice. In T. Constantion and B. White (Eds.), *Essays on aesthetic education for the 21st century* (pp. 81-92). Rotterdam: Sense.

Stanford Encyclopedia of Philosophy. (n.d.) Peirce's theory of signs. Retrieved from http://plato.stanford.edu/entries/peirce-semiotics/

Stafleu, F. (1968). Troll's roots and root systems. *Taxon, 17*(1), 73-75.

Stenberg, G., Radeborg, K., & Hedman, L. R. (1995). The picture superiority effect in a cross-modality recognition task. *Memory & Cognition, 23*(4), 425-441.

Stewart, M. (2014). Enduring understandings, artistic processes, and the new visual arts standards: A close-up consideration for curriculum planning. *Art Education, 67*(5), 6-11.

Taber, K, & Watts, M. (1996). The secret life of the chemical bond: students'anthropomorphic and animistic references to bonding. *International Journal of Science Education, 18* (5), 557-568.

Tang, K. S., & Moje, E. B. (2010). Relating multimodal representations to the literacies of science. *Research in Science Education, 40*(1), 81-85.

Taylor, E. W. (1964). Nature in art in Renaissance literature. New York: Columbia College Press.

Tharp, R. G., Doherty, R. W., Echevarria, J., Estrada, P., Goldenberg, C., Hilberg, R. S., et al. (March 2004). *Five Standards for Effective Pedagogy and Student Outcomes* (No. G1). Berkeley, CA: University of California, Berkeley. Retrieved from http://crede.berkeley.edu/research/crede/products/print/occreports/g1.html

Tochon, F. (2013). *Educational semiotics: Signs and symbols in education*. Blue Mounds, WI: Deep University Press.

Trefil, J. (2008). *Why Science?* New York: Teacher's College.

Turner, V. W. (1982). Performing ethnography. *Drama Review, 23*, 33-55.

Tytler, R., Prain, V., Hubber, P, & Waldrip, B. (Eds.). *Constructing representations to learn in science*. Rotterdam, The Netherlands: Sense.

Uhrmacher, P.B. (2010). The power to transform: Implementation as aesthetic awakening. In T. Constantion & B. White (Eds.), *Essays on aesthetic education for the 21st century* (pp. 183-203). Rotterdam: Sense.

U.S. Department of Education, Institute of Education Sciences, National Center for Education Statistics, National Assessment of Educational Progress (NAEP), 2009 and 2011 Science Assessments.Grade 8 State Results. Retrieved http://nationsreportcard.gov/science_2011/g8_state.asp?subtab_id=Tab_4&tab_id=tab1#chart

Van Driel, J. H., Beijarrd, D. & Verloop, N. (2001). Professional development and reform in science education: The role of teachers' practical knowledge. *Journal of Research in Science Education, 38*(2), 137-158.

Van Sickle, M., & Spector, B. (1996). Caring relationships in science classrooms: A symbolic interaction study. *Journal of research in science teaching, 33*(4), 433-453.

Wallace, R. (1966). *The world of Leonardo: 1492-1519.* Alexandrai, VA: Time Life.

Windschitl, M., Thompson, J., & Braaten, M. (2008). Beyond the scientific method: Model-based inquiry as a new paradigm of preference for school science investigations. *Science Education, 92*(5), 941-967.

Windshitl, M. et al (2013). The big idea tool. Retrieved from http://tools4teachingscience.org/tools/big_idea.html

Winerman, L. (2005). The mind's mirror. *Monitor on psychology, 36*(9). Retrieved from http://155.97.32.9/~bbenham/Minds%20and%20Morals/The%20mind%27s%20mirror.APA.pdf

Wisely, F. (1994). Communication Models. In D. Moore & F. Dwyer (Eds.), *Visual literacy: A spectrum of visual learning.* Englewood Cliffs, NJ: Educational Technology.

DEEP UNIVERSITY PRESS
SCIENTIFIC BOARD MEMBERS

Dr. Araceli Alonso, Global Health Institute, Department of Gender and Women's Studies, University of Wisconsin-Madison, USA

Dr. Ronald C. Arnett, Chair and Professor, Department of Communication & Rhetorical Studies, Duquesne University

Dr. Gilles Baillat, Rector, ex-Director of CDIUFM Conference of French Teacher Education Directors, University of Reims, France

Dr. Niels Brouwer, Graduate School of Education, Radboud Universiteit Nijmegen, The Netherlands

Dr. Jianlin Chen, Shanghai International Studies University, China

Dr. Yuangshan Chuang, President, APAMALL, NETPAW, Taiwan

Dr. Enrique Correa Molina, Professor and Vice-Dean, Faculty of Education, University of Sherbrooke, Canada

Dr. José Correia, Dean of Education, University of Porto, Portugal

Dr. W. John Coletta , Professor, Univ of Wisconsin-Stevens Point

Marc Durand, Professor, University of Geneva, Switzerland

Dr. Paul Durning, Doctoral School, French National Observatory, EUSARF, University of Paris X Nanterre, Paris, France

Dr. Manuel Fernandez Cruz, Professor, University of Granada, Spain

Dr. Stephanie Fonvielle, Associate Professor, Teacher Education University Institute, University of Aix-Marseille, France

Dr. Elliot Gaines, Professor, Wright State University, President of the Semiotic Society of America, Internat. Communicology Institute

Dr. Mingle Gao, Dean, College of Education, Beijing Language and Culture University (BLCU), Beijing, China

Dr. Mercedes González Sanmamed, Professor, Univ of Coruña, Spain

Dr. Gabriela Hernández Vega, Professor, Univ of Nariño, Colombia

Dr. Teresa Langle de Paz, Autonomous University, Feminist Research Institute Council, Complutense University of Madrid, Spain

Dr. Maria Masucci, Drew University, New Jersey, USA

Dr. Joëlle Morrissette, Professor, Department of Educational Psychology, Université of Montreal, Quebec, Canada

Dr. Martha Murzi Vivas, Professor, Univ of Los Andes, Venezuela

Dr. Thi Cuc Phuong Nguyen, Vice Rector, Hanoi University, Vietnam

Dr. Shirley O'Neill, Associate Professor, President of the International Society for leadership in Pedagogies and Learning, University of Southern Queensland, Australia

Dr. José-Luis Ortega, Professor, Foreign Language Education, Faculty of Education, University of Granada, Spain

Dr. Surendra Pathak, Head and Professor, Department of Value Education, IASE University of Gandhi Viday Mandir, India

Dr. Luis Porta Vázquez, Professor at the National University of Mar del Plata CONICET (Argentina)

Dr. Shen Qi, Associate Professor, Shanghai Foreign Studies University (SHISU), Shanghai, China

Dr. Timothy Reagan, Professor and Dean of the College of Education at Zayed University in Abu Dhabi/Dubai, Saudi Arabia

Dr. Farouk Y. Seif, Exec. Director of the Semiotic Society of America, Center for Creative Change, Antioch University Seattle, Washington

Dr. Gary Shank, Professor, Educational Foundations and Leadership, Duquesne University, Pittsburgh, Pennsylvania

Dr. Kemal Silay, Professor, Flagship Program Director, Department of Central Eurasia, Indiana University-Bloomington, USA

Dr. José Tejada Fernández, Professor at the Autonomous University of Barcelona, Spain

Dr. François Victor Tochon, Professor, University of Wisconsin-Madison, USA

Dr. Brooke Williams Deely, Women, Culture and Society Program, Philosophy Department, University of St. Thomas, Houston

Portable Digital Microscope
ATLAS OF CERAMIC PASTES
COMPONENTS, TEXTURE AND TECHNOLOGY
Isabelle C. Druc

**with the technical collaboration of
Bruce Velde and Lisenia Chavez**

This manual is the first of its sort describing the use of the new portable digital microscope for analysis of archaeological ceramics in the field or in the laboratory. It is presented like a geological atlas with a description of the most common minerals and lithic fragments found in ancient ceramic pastes to help archaeologists identify what they see under the microscope. Identification of manufacture and technological features are also addressed. An analytic protocol is proposed along with further suggestions for granulometric and digital image analyses to help with the constitution of groups of similar composition and paste texture. The manual is abundantly illustrated with pictures of archaeological and ethnographic ceramic pastes and raw materials. It is a reference book for all involved in the analysis of archaeological ceramics and a major tool to help study, classify and choose the best fragments for archaeometric analysis.

> This timely and valuable contribution led by Dr Isabelle Druc, a renowned ceramic specialist, brings the spotlight back to the study of pottery and its myriad relationships with people. This handy guide will be useful for both students and professionals interested in learning to investigate, with precision, the composition of raw materials and their transformation by people. It enables the identification and description of their choices that inform about the critical stuff (techniques, identity, values, landscape) of ancient cultures.
>
> **—George Lau, University of East Anglia, UK**
>
> Given the increased accessibility of tools such as portable microscopes, this book provides timely and very useful guidelines for macroscopic analysis of ceramic paste. With its detailed illustrations, descriptions of diagnostic features for different kinds of minerals, and holistic approach to systems of ceramic production, I believe the book will be regarded as essential to consult for initial research on paste composition in many areas.
>
> **—Anne Underhill, Yale University**

http://www.deepuniversitypress.org/atlas.html

PERFORMING THE ART OF
LANGUAGE LEARNING
Deepening the Learning Experience through Theatre and Drama

Dr. Kelly Kingsbury Brunetto
University of Nebraska-Lincoln, USA

Truly innovative, *Performing the Art of Language Learning* delivers an exhaustive account of the role theater can and should play in second language acquisition. Kingsbury-Brunetto makes a compelling case for the integration of the performing arts within foreign language and literature departments. This will surely be an influential study for the advancement of the field.
—*Florent Masse, Director, L'Avant-Scène,*
The French Theater Workshop, Princeton University, U.S.A.

This is a well-researched and beautifully written text investigating how engagement with theater in courses designed for language acquisition and development can enhance undergraduate university students' learning. Grounded in Bakhtinian notions regarding discourse practices and Van Lier's ecological approach to second language acquisition, Professor Kingsbury Brunetto has produced a theory-rich book that also is highly readable and enjoyable. The text is methodologically rigorous and rich in detail concerning students' understandings and interactions with one another, their faculty members, the plays they enacted, and their audiences. Also included after each chapter are questions for readers' critical reflection that should produce complex discussions among readers, and especially will be helpful in graduate classes in both second language acquisition and theater.
——*Mary Louise Gomez, Professor, Languages and Literacies,*
Teacher Education, University of Wisconsin-Madison, U.S.A

I find the book very inspiring and valuable. I have been using drama and theatre in language courses for fifteen years and I still continue to expand my comprehension of their enormous potential for learning. Dr.Kingsbury Brunetto's thoroughly crafted work is a much appreciated addition to my growing understanding of the manifold processes that make the learning happen. We absolutely need research projects like this one to help drama and theatre assume a more central position in the language teaching world.
——*Barbora Müller Dočkalová, Faculty of Education, Charles*
University in Prague, Czech Republic

http://www.deepuniversitypress.org/performing.html

SIGNS AND SYMBOLS IN EDUCATION
EDUCATIONAL SEMIOTICS

François Victor Tochon, Ph.D.
University of Wisconsin-Madison, USA

In this monograph on Educational Semiotics, Francois Tochon (along with a number of research colleagues) has produced a work that is truly groundbreaking on a number of fronts. First of all, in his concise but brilliant introductory comments, Tochon clearly debunks the potential notion that semiotics might provide yet another methodological tool in the toolkit of educational researchers. Drawing skillfully on the work of Peirce, Deely, Sebeok, Merrell, and others, Tochon shows us just how fundamentally different semiotic research can be when compared to the modes and techniques that have dominated educational research for many decades. That is, he points out how semiotic methods can provide the capability for both students and researchers to look at this basic and fundamental human process in inescapably transformational ways, by acknowledging and accepting that the path to knowledge is, in his words "through the fixation of belief."

But he does not stop there – instead, in four brilliantly conceived studies, he shows us how semiotic concepts in general, and semiotic mapping in particular, can allow both student teachers and researchers alike insights in these students' development of insights and concepts into the very heart of the teaching and learning process. By tackling both theoretical and practical research considerations, Tochon has provided the rest of us the beginnings of a blueprint that, if adopted, can push educational research out of (in the words of Deely) its entrenchment in the Age of Ideas into the new and exciting frontiers of the Age of Signs.

Gary Shank
Duquesne University

SEE REVIEWS HERE: http://www.deepuniversity.net/book1.html

Guide to Authors

What our Publishing Team can offer:

- An international editorial team, in more than 20 universities around the world.

- Dedicated and experienced topic editors who will review and provide feedback on your initial proposal.

- A specific format that will speed up the production of your book and its publication.

- Higher royalties than most publishers and a discount on batch orders.

- Global distribution and marketing through Amazon and Barnes & Noble in the U.S., UK, Australia, Europe, China, and many other countries.

- Fair recognition of your work in your area of specialization.

- Quality design. Using the latest technology, our books are produced efficiently, quickly and attractively.

- A global marketing plan, including electronic and web marketing and review mailing.

- Book Series: Deep Education; Deep Language Learning; Signs & Symbols in Education; Language Education Policy; Deep Professional Development; Deep Early Childhood Education; Deep Activism.

http://www.deepuniversitypress.com/universitypress.html

Contact : publisher@deepuniversity.net

Deep University Online !

For updates and more resources
Visit the Deep University Website:

www.deepuniversity.net

www.depuniversitypress.org

Contact : publisher@deepuniversity.net

❖ Online Certificate and cMOOCs on Deep Education: http://www.deepuniversity.net/graduatecourses.html

❖ Facebook group on Signs & Symbols in Education : https://www.facebook.com/groups/EducationalSemiotics/

Correspondence with the author:

merriekoester@comcast.net

Merrie Koester's Biosketch

Merrie Koester, Ph.D., science educator and author of the nationally implemented *Agnes Pflumm* science education novels, has been bringing students to science through literature and the creative arts for the last 25 years. A native of Charleston, SC, adjunct professor in science education at the College of Charleston, and a Phi Beta Kappa graduate of Furman University, Dr. Koester first began developing curriculum for teaching science through the arts as part of her masters' research at the University of Hawaii in 1990. As a science and arts integration specialist, she is dedicated to working with each state, district, school and individual teacher to improve achievement in STEAM education and has facilitated professional development workshops on teaching science through the arts, speaking at district, state, and national level science education conferences. A key feature of Dr. Koester's curriculum is the deepening of science pedagogical content knowledge through the practice of what she calls performative narrative drawing and the creation of "Know"tations. This book is derived from her doctoral dissertation, *Science Teachers Who Draw: A Metasemiotic Curriculum Inquiry*, at the University of South Carolina. As part of her doctorate, she developed *Project Draw for Science*, a collaborative action research initiative in South Carolina for communicating meaning in science through drawing. *Project Draw for Science* is now a new program of study and professional development initiative for the University of South Carolina Center for Science Education. She hopes that this book will inspire her readers to celebrate the artist within themselves.

❖ Merrie Koester's Facebook Page :

https://www.facebook.com/scienceteacherswhodraw